IDAHO SOILS ATLAS

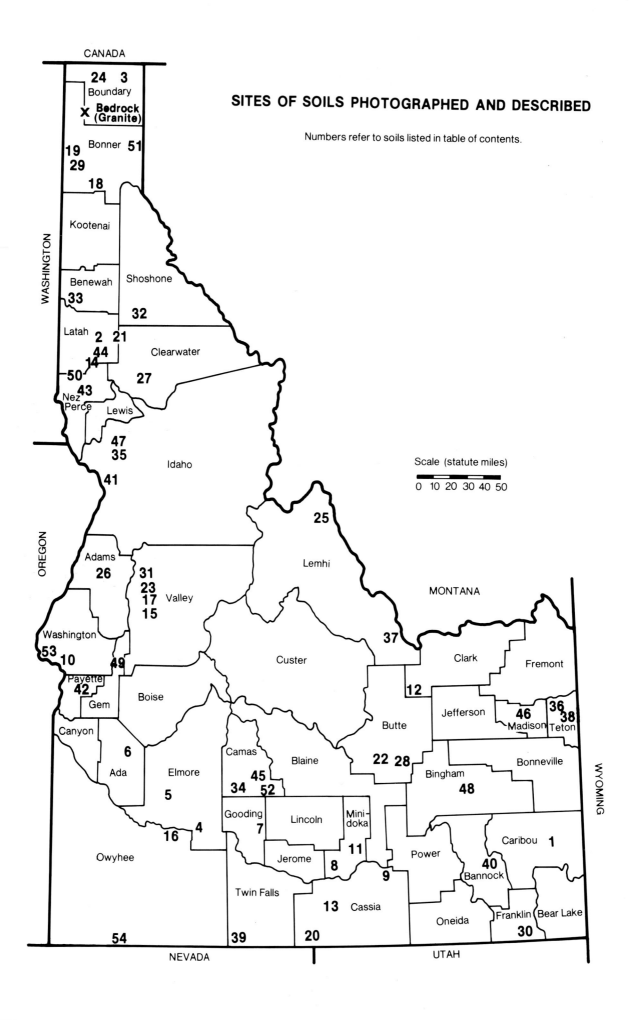

Idaho Soils Atlas

by
Raymond J. Barker
Robert E. McDole
Glen H. Logan

Photography
by
Raymond J. Barker

The University Press of Idaho
A Division of
The Idaho Research Foundation, Inc.
1983

ISBN 0-98301-088-X
Library of Congress Catalog Card number 82-60201
Copyright © 1983 by the University Press of Idaho
All rights reserved
Manufactured in the United States of America
Printed by the
News-Review Publishing Co., Inc.
Moscow, Idaho 83843

Cover design RMSB - Liz Mowrey
Published by the University Press of Idaho
A Division of the Idaho Research Foundation, Inc.
Box 3368, University Station, Moscow, Idaho 83843

CONTENTS

Map showing sites of soils
    photographed and described .... Frontispiece
Acknowledgments 4
Introduction 5
Soil development 7
    Parent material 7
    Climate 7
    Living organisms 8
    Topography 9
    Time 9
Key to prominent soil characteristics 10
Soil 1 .... **12**
Soil 2 (Santa Series) **14**
Soil 3 (Porthill Series) **16**
Soil 4 (Chilcott Series) **18**
Soil 5 (Colthorp Series) **20**
Soil 6 (Sebree Series) **22**
Soil 7 (Gooding Series) **24**
Soil 8 (Portneuf Series) **26**
Soil 9 (Trevino Series) **28**
Soil 10 (Owyhee Series) **30**
Soil 11 (Minidoka Series) **32**
Soil 12 .... **34**
Soil 13 (Garbutt Series) **36**
Soil 14 (Flybow Series) **38**
Soil 15 (Pyle Series) **40**
Soil 16 (Quincy Series) **42**
Soil 17 (Shellrock Series) **44**
Soil 18 (Pywell Series) **46**
Soil 19 (Vay Series) **48**
Soil 20 (Bluehill Series) **50**
Soil 21 .... **52**
Soil 22 (Moonville Series) **54**
Soil 23 (Roseberry Series) **56**
Soil 24 .... **58**
Soil 25 .... **60**
Soil 26 .... **62**
Soil 27 .... **64**
Soil 28 (Moonville Variant) **66**
Soil 29 (Bonner Series) **68**
Soil 30 (Oxford Series) **70**
Soil 31 (McCall Series) **72**
Soil 32 .... **74**
Soil 33 (Southwick Series) **76**
Soil 34 (Houk Series) **78**
Soil 35 (Nez Perce Series) **80**
Soil 36 (Driggs Variant) **82**
Soil 37 .... **84**
Soil 38 (Greys Series) **86**
Soil 39 .... **88**
Soil 40 (Pavohroo Series) **90**
Soil 41 (Tannahill Series) **92**
Soil 42 (Gem Series) **94**
Soil 43 (Gwin Series) **96**
Soil 44 (Klickson Series) **98**
Soil 45 (Little Wood Series) **100**
Soil 46 (Rexburg Series) **102**
Soil 47 (Westlake Series) **104**
Soil 48 (Hymas Series) **106**
Soil 49 (Ola Series) **108**
Soil 50 (Palouse Series) **110**
Soil 51 .... **112**
Soil 52 (Magic Series) **114**
Soil 53 (Ager Series) **116**
Soil 54 (Boulder Lake Series) **118**
Bedrock (Granite) **120**
Literature Cited 122
Appendix 123
Glossary 123
    Soil classification (Taxonomy) 129
    Depth to layer impeding root
        development 131
    Natural soil drainage 132
    Parent material 133
    Average annual precipitation 135
    Average annual air temperature 137
    Average frost-free season 139
    Elevation 141
    Topography 143
    Habitat type 144
    List of plant names 146
    Land use 147

ACKNOWLEDGMENTS

This book is the result of a joint effort of the University of Idaho, College of Agriculture; the Soil Conservation Service and Forest Service of the United States Department of Agriculture; the Bureau of Land Management, Bureau of Reclamation, and Bureau of Indian Affairs of the United States Department of Interior; and the Idaho Soil Conservation Commission. Each of these agencies was involved in the overall planning of the Idaho Soils Atlas. They also furnished technical data and financial support.

Further contributing financial support includes BN Timberlands, Inc. and Idaho Environmental Health Association.

Other groups having a strong interest in the project include the R.N. Irving Chapter of the Soil Conservation Society of America and the Idaho State Supervisor of Agricultural Education of the State Board for Vocational Education.

Special thanks are given to the many soil scientists throughout Idaho who worked in the field locating typical sites of the soils which were chosen for the Idaho Soils Atlas. They also were involved in preparing the soil profiles for photographing, describing the soils, and sampling the many soil horizons for later laboratory studies by the University of Idaho. Special thanks to Dr. M. A. Fosberg and his staff for their assistance in locating, describing, and sampling sites and cataloging and storing soil samples collected at each site.

INTRODUCTION

This *Idaho Soils Atlas,* featuring 54 soils from throughout the state, has been prepared as an aid to further the knowledge of soils.

The objective of this atlas is to present all of the prominent soil characteristics or properties which exist in the state and to show how these characteristics affect use and management. Properties include physical as well as chemical. This atlas is meant to be only an introduction to understanding the complex nature of soils in Idaho. It is also an introduction to more complete information available in soil survey publications.

The soils included in this atlas were selected from several hundred different soils which have been identified in the state. Each of the soils in the state has characteristics that make it a unique individual soil. The *Idaho Soils Atlas* includes soils showing all of the prominent soil characteristics. Thus, prominent characteristics of any of the several hundred soils in Idaho can be found among the 54 soils included in this atlas. For example, a soil having a thin dark-colored surface, a well developed subsoil, and shallow depth to bedrock is not included in the atlas, but all of these characteristics can be found in the soils which are included.

It was not only impossible but impractical to treat all the soils which occur in the state. Instead, soils were selected to show the full range of prominent properties which exist within the state. There are deep soils and shallow soils; clayey, silty, and sandy soils; poorly drained soils and well drained soils; weakly developed and well developed soils; warm soils and cool soils; light colored and dark colored soils; soils with and without rock fragments; soils developed in all the major geologic sources of material; nearly level soils and steep soils; soils in arid climates and soils in moist climates; acid soils and alkaline soils; and many more. All soil characteristics shown in the atlas also occur in soils outside the state.

Photographs of the soils show the various layers or soil horizons. Along the left hand side of the soil profile photos are depths, in feet, from the land surface. Marks on the right hand side show boundaries of the major soil horizons. The horizons have been identified by letters and numerals (3) which are used by soil scientists to show the soil forming process or processes which they believe have taken place in the development of each soil horizon. These horizon designations make possible useful comparisons among soils. All A horizons are not alike nor are all B horizons alike. They only indicate the observers' considered judgment as to the kinds of departure from the original parent material. The glossary briefly describes these letters and numerals.

Each soil horizon is briefly described. Soil color names are from Munsell color charts. If the soils are normally dry during summer months then the dry colors are given. For soils which are mostly moist, such as poorly drained or somewhat poorly drained soils, moist colors are given. Most soils become lighter colored when they dry.

The glossary may be helpful to understand certain words used in the descriptions given for each soil horizon. All of the soil descriptions have been greatly simplified. In many cases similar soil horizons have been combined. More detailed information is available for these soils in soil survey publications.

All soils occur in a particular landscape or topography. For example, **Soil 34 (Houk Series)** is on nearly level bottom lands or low terraces, **Soil 50 (Palouse Series)** is on gently sloping to hilly uplands, and **Soil 41 (Tannahill Series)** is on steep and very steep south and west-facing slopes of canyons. The landscape photograph shown with each soil is typical for that soil. In many places other soils are shown nearby that are different in some respects. The major differences of these soils are briefly described.

On the page opposite the photographs is a discussion of each soil. Most of the soils have been given a name called a series, like Santa Series, for easy reference, like trees and flowers are named as species. When a soil has been identified, described, classified, and named, it is recognized throughout the United States as that individual. No other soil can use that name. These names are taken mainly from local towns, counties, lakes, and streams. Each soil series has either different kinds of soil horizons or significantly different properties in one or more soil horizons.

Soil variants are soils which are somewhat similar to the named soil series, but their interpretations for use and management are significantly different. Soil variants have limited geographic areas which do not justify creation of a new series.

All soils are classified according to a nationwide system (1). This classification system has been developed to classify the many thousands of soils in the United States. It resembles the systems used for classifying plants and animals. The soil classification system is used mainly by soil scientists to group soils having similar properties, compare one soil to another, and as an aid to identifying a given soil.

Soil classification at the family level category is followed by several items which characterize each soil: soil depth, drainage class, parent material, climate, elevation, topography, habitat type, common native vegetation (2), the general location of the soil in the state, and the dominant land use. This helps to visualize the setting of each soil.

The next section, entitled **"The Prominent Characteristics of This Soil Are"**, gives a few main properties of each soil that affect use and management. Many more properties could be listed but these are dominant. The selection of soils to be included in this atlas was made because they show a large variety of soil properties.

Finally on that page is a discussion of the soils including the possible soil forming processes active in developing the soils, especially the prominent characteristics listed for each soil, and their affect on use and management. Not all soil properties nor all soil interpretations are given for the soils. The intent here is to show the wide variations of soil properties occurring in the state that result in a wide range of uses and management choices. More complete information for these and many other soils are available in soil survey publications.

NOTE: Colors are for dry soils in the profile photographs and descriptions, with the exception of Soils 18, 21, 23, 29, 34,, 47, and 51 which depict moist soils.

SOIL DEVELOPMENT

A new concept of soils was developed a little more than a hundred years ago by a small group of scientists in Russia, under the leadership of Dokuchaiev. They noted the differing soil horizons and came to the conclusion that a whole soil, from the surface to the lower depths, developed naturally by a unique combination of soil-forming factors. These are the earthy material or *parent material* that was altered by the effects of *climate, living organisms,* and *topography,* over a period of *time.* These five factors are the genetic factors which are directly and wholly responsible for the characteristics and qualities of each soil individual. These genetic factors of soil formation are discussed separately.

PARENT MATERIAL

Geological events over a great period of time have provided Idaho with many different kinds of parent material. Each kind has contributed different characteristics to soils, physically as well as chemically. In many cases, soils have developed in two or more kinds of material, one on top of the other or mixed together.

There are two general kinds of mineral parent materials — those that weathered in place from old rock formations (residuum) and those that were transported by wind, water, or ice.

There are many rock formations in Idaho. They include granite, gneiss, schist, limestone, and basalt. These rock formations are most easily seen in the mountainous parts of the state. Basalt occurs in large plateaus and plains in the northern part of the state and in the Snake River plain of the southern part. All of these rocks have undergone various degrees of weathering.

Parent materials moved by wind include sands and silts. Sands moved near the ground surface for relatively short distances. They appear mostly as dunes in rather small areas in southern Idaho along the Snake River and near Bonners Ferry in northern Idaho. Silt, with a small amount of very fine sand and clay, was carried longer distances. This is known as loess. Deposits of loess are extensive in northern Idaho, mostly on the Columbia River basalt plateaus, as well as large areas in the southern part of the state. In more moist areas some of the silts in the loess weathered to clay.

Volcanic ash of various ages, from recent times (a few hundred years ago) to quite old (at least 3 million years), was also carried by wind and deposited as an important parent material in various parts of the state. Volcanic ash contrasts with most parent materials in being light weight.

Water has moved great quantities of material in all parts of the state, providing a wide variety of parent materials for soils. They grade from coarse materials, like sand, gravel, and cobblestones, where the water has moved fast to very fine materials, like silt and clay, where the water was still or moved slowly. These materials came from a wide variety of rock formations which have influenced the physical and chemical properties of the soils.

Glaciers provided yet another kind of mineral parent material. Large glaciers advanced several times into Idaho from Canada as far south as Coeur d'Alene during the Pleistocene age. Small glaciers also occurred locally in the higher mountains throughout the state. A mixture of various sizes of rock along with finer materials was deposited by the ice.

Peat is a special kind of parent material that forms into organic soils. Organic soils mostly occur in bogs north of Coeur d'Alene and in parts of eastern Idaho. They may contain minor amounts of mineral material deposited by floods or wind.

These parent materials were subject to the other factors of soil formation — climate and living organisms modified by topography over a period of time. Actually, all the soil forming factors are working at the same time with different rates of speed or efficiency. They constantly interact with one another.

CLIMATE

Even a casual observer readily notes that there are many contrasting climates in Idaho. Precipitation ranges from about 8 inches per year with very little snow along the Snake River south of Boise, to over 60 inches with many feet of snowfall in the higher elevations of northern Idaho. Great changes of temperature and frost-free days per year also occur.

In some places, the climate changes within very short distances. This is especially true with an abrupt rise in elevation in the mountains. Also the climate of a steep north slope can differ markedly from a nearby steep south slope. This is most noticeable in the canyons along the Snake and Salmon Rivers where north-facing slopes are more moist and cooler than south-facing slopes.

What is it about climate that affects the formation and character of soils? There are direct as well as indirect effects. The amount of water that moves into and through a soil affects the rate at which weathering takes place. In the total absence of water, such as on the moon, this simply doesn't happen.

Water puts into solution many materials, such as compounds of potassium, sodium, calcium, magnesium, and silicon. The more water available, especially in a soil material that allows water to move easily, the more these compounds are removed from the soil. This can be seen in several of the soil profile pictures. Calcium carbonate is light colored. In soils having limited rainfall it has been moved by water from surface soil horizons into deeper layers. This is a common feature of many soils in the Snake River plain area and near Lewiston. With more annual precipitation, calcium carbonate is washed completely out of the soil.

Different chemical changes take place within soils influenced by the amount of water present and the soil temperature. With a state as varied as Idaho in precipitation and temperature, there are wide differences in soils. In general the drier parts of the state have higher ranges of soil reaction or pH. Most of these soils are neutral to strongly alkaline. In the more moist areas, pH values range from neutral to strongly acid. Availability of most plant nutrients is generally related to pH. Extremes of acidity or alkalinity adversely affect this availability.

Soils with an excess of moisture like those with fluctuating water tables are affected in yet another way. They are nearly always mottled with various shades of gray, brown, and yellow. Soils with mottles are saturated with water at some time of the year. Such soils in Idaho are usually in low-lying positions.

Water is also responsible for moving clay downward into lower soil horizons. Clay that has lodged in these horizons can be seen by using a hand lens. It appears as clay films which resemble candle wax drippings. All the soil horizons having a lower case "t", such as Bt, have these clay films.

Chemical processes occur much more rapidly under high temperatures than under low temperatures. Alternate wetting and drying of soils can also affect chemical changes as well as physically mix soil materials, especially soils high in clay. In addition, freezing and thawing mixes soil materials.

Climate has an indirect effect on soil development. It influences to a large extent the kind and amount of vegetation.

LIVING ORGANISMS

Plants and animals, including man, affect the natural development of soils. Different kinds of vegetation incorporate varying amounts and kinds of residue into the soil. This residue of leaves, needles, stems, and roots are converted to organic matter by the many microorganisms present in the soil. The soil horizons most affected are near the surface where most of the residue occurs. Generally in the temperate climate of Idaho, dark colored soils have more organic matter than light colored soils.

There is a wide assortment of natural plant communities or habitat types, in Idaho. In the driest and warmest part south of Boise, the vegetation is relatively sparse. The natural level of organic matter is low — well below 1 percent. The soils are light colored. With a rise in precipitation and decrease in temperature, the vegetation becomes more abundant, and the organic matter in the soils increases. This reaches the highest level under dense grasses that once covered the prairie near Grangeville. Organic matter in the black soils of this area was nearly 10 percent.

With further increases in precipitation and cooler temperatures, coniferous forests begin. There are fewer grasses and forbs under the trees which provide a low amount of organic material. Soils are more strongly leached, lighter colored, and contain less organic matter. At higher elevations, however, the surface soil horizons are darker colored and are high in organic matter. Colder temperatures at these elevations retard the decomposition of organic matter.

Microorganisms are present in all soils in varying amounts. When organic residue from plants and animals is provided, their numbers rise spectacularly. These tiny organisms are largely responsible for the decomposition of organic material.

Larger organisms are also present such as insects, worms, rodents, and other burrowing animals. They contribute to the decomposition of organic material as well as generally mix the soil.

Man has contributed to the present nature of soils in dramatic ways. Clearing the forests for cultivation has permanently changed the natural soil temperature and the rate of water runoff.

Breaking out the natural prairies and clearing the forests for cultivation has mixed the surface soil and has in many places allowed great quantities of soil to be eroded. Soils are no longer as deep and the surface soils are lower in organic matter and have poor tilth.

Many soils in the Snake River plain of southern Idaho have been leveled for more efficient surface irrigation. Some soils have been cut (soil removed) while others have been filled (soil deposited).

TOPOGRAPHY

Wide variations of geologic events in Idaho have resulted in many different landscapes. Topography, or lay of the land, is an important soil forming factor. It affects the natural soil erosion as well as accumulations of alluvium over long periods of time. It also affects the occasional rapid movement of soil down slopes, the natural soil drainage, air drainage, and even the local climate.

Nearly level valley floors have slow or very slow runoff and are generally moderately well drained to poorly drained. Little soil erosion takes place. Ratner, soil material is generally added by periodic flooding. These positions also have lower temperatures than adjacent uplands due to cold air drainage.

As the slope increases, there are progressive changes. Runoff is increased. Soil is lost by erosion rather than being accumulated. The soils become better drained. Air drainage is improved. All of these things are intensified with continued steepening of the slope. Soils on very steep slopes erode not only by water runoff, but by soil actually creeping or rapidly sliding downslope. Larger fragments of gravel, cobblestones, and stones periodically roll down hill. Soils on very steep slopes commonly are either very shallow or contain a high percentage of rock fragments.

The direction in which slopes face in Idaho's latitudes affects the local climate. This in turn affects soil development. Soils with south-facing slopes are significantly warmer and dry out faster than soils on nearby north-facing slopes. Strongly contrasting plant communities exist on north slopes compared to south slopes.

The particular shape and position of slopes also affects water runoff. Convex slopes contrast with concave slopes. Convex ridge positions disperse water more uniformly than do concave sloping areas which concentrate water causing greater runoff and erosion.

TIME

Last, but equally important, is the effect of time on the formation of soils. Time is required for all things to happen. Time is merely relative, however, when understanding its effect on soil development. It is not just the number of years that has passed but the intensity of the soil forming processes during a given period of time that determines soil development. For example, a parent material available for soil development in a dry cold climate with sparse vegetation would develop exceedingly slow. In fact, not much would ever happen. The same kind of parent material in a warm moist climate with abundant vegetation would develop at a much faster rate.

So the question which is often asked about how long does it take to make an inch of topsoil has many answers. The minimum time required is believed to be many hundreds of years.

Development of a soil generally reflects its age. A soil is considered to be "young" if its soil horizons are weakly expressed. The soil may have some accumulation of organic matter in the surface layer and only very weak development of a B horizon or none at all. A soil is "old" if its soil horizons are well developed. An example of an "old" or older soil would be one showing evidence of clay movement from a surface horizon into the subsoil.

Now it is much easier to understand why there are so many different soils in Idaho, even within each county. There is seemingly an endless variety of significantly different combinations of the five soil forming factors. It is the combined effects of a particular set of these genetic factors that result in a "soil".

KEY TO PROMINENT SOIL CHARACTERISTICS

The following key may be used as an aid in locating soils in the atlas which were chosen to represent the more prominent soil characteristics affecting use and management. These have been arranged for ease in finding those properties of interest.

For complete listings of soil depth, drainage, and parent material refer to tables on pages 130, 131, and 132.

Pages

AVAILABLE WATER HOLDING CAPACITY
- High 53, 67, 111
- Low 29, 43, 61, 69
- Very Low 41, 45, 73

DEPTH TO RESTRICTIVE LAYER
- Very deep 111
- Moderately deep to bedrock 21, 95
- Moderately deep to fragipan 15
- Shallow to bedrock 29, 97, 107
- Shallow to duripan 21, 35
- Very shallow to bedrock 39

SOIL DEVELOPMENT
- Acid reaction 57, 75
- Clayey soil 115, 117, 119
- Clayey subsoil with high shrink-swell 71, 81, 95
- Duripan formed in coarse fragments 35
- Duripan, strongly cemented 19, 23, 33, 89
- E soil horizon(s) at the surface 13, 17, 23
- E soil horizon below the surface soil 15, 77, 79, 81, 87
- Grayish colored soil horizons 79, 119
- High organic matter in surface soil 81
- Leached upper subsoil 31
- Light colored surface soil 19, 25, 27, 33, 35, 37
- Lime accumulation below the surface soil 27
- Lime accumulation deep in the soil 25, 31, 85, 93, 103
- No soil development 43
- No subsoil development 37, 45, 109
- Reddish colored subsoil 63, 113
- Sodium in the subsoil 23
- Thick, very dark colored surface layer 57, 85, 91, 105, 109
- Thick, dark colored surface layer 111
- Thin, dark colored surface layer 61
- Weakly developed subsoil 51, 61, 63, 103, 107
- Well developed subsoil 13, 19, 25, 89, 95, 97
- Wide cracks to the surface when dry 115, 117, 119

SOIL DRAINAGE
- Flooding hazard 79, 105
- Fluctuating water table 57, 79, 105
- Very poorly drained 47

SOIL MATERIAL
- Apparent coarse fragments which easily break down 51
- Boulders on the surface 59, 73

Cinders in substratum 55
Easily rippable bedrock 41, 45, 65, 109
Gravel and sand substratum 69, 83, 101
Highly erodible surface soil... 15
Highly erodible volcanic ash surface soil 49, 51, 53, 55, 59, 65, 67, 75, 113
Laminated substratum 31
More than 35 percent coarse fragments 49, 61, 93, 97, 99, 101
More than 35 percent rock fragments 59, 73
Organic soil 47
Sandy textures 41, 43, 45, 57
Stones on the surface 115

SOIL PERMEABILITY
Very slow... 15, 25, 115, 117
Slow 17, 31, 77
Moderate 103
Rapid 43, 55

OTHER
Barren soil under natural conditions 23
Intermound soil 39
Short frost-free season 79, 83, 85, 87, 91
Steep slopes 65, 75, 93, 99
Susceptible to frost action 87, 111

Soil 1*

*Colors in photograph and description are for dry soil in this and subsequent profiles, except for soils 18, 21, 23, 29, 34, 47, and 51 which are for moist soil.

E1—0 to 5 inches; pale brown silt loam, granular structure, strongly acid.

E2—5 to 15 inches; pale brown silt loam, subangular blocky structure, strongly acid.

BE—15 to 26 inches; yellowish brown and very pale brown silty clay loam, angular blocky structure, few gravel, strongly acid.

Bt1—26 to 43 inches; brownish yellow and yellowish brown silty clay loam, prismatic and angular blocky structure, few gravel, strongly acid.

Bt2—43 to 53 inches; yellowish brown very gravelly silty clay loam, subangular blocky structure, violently effervescent, neutral.

Bt3—53 to 61 inches; light yellowish brown very gravelly silty clay loam, subangular blocky structure, violently effervescent, neutral.

Soil 1 is on wooded slopes in the background. Soils in the foreground, under a mixture of shrubs and grasses, have dark colored surface layers.

Soil 1

Family Classification: fine-loamy, mixed, Typic Cryoboralfs
Soil Depth: 60 inches or more
Drainage Class: well drained
Parent Material: loess over limestone residuum
Average Annual Precipitation: about 25 inches
Average Annual Air Temperature: about 38 degrees F.
Average Frost-Free Season: 50 days or less
Elevation: 6,600 to 8,000 feet
Topography: moderately sloping to steep foothills and mountains
Habitat Type: *Abies lasiocarpa/Pachystima myrsinites*
Common Native Vegetation: subalpine fir, Douglas-fir, lodgepole pine, myrtle pachystima, snowberry, rose, willow, pine reedgrass, and geranium
Occurrence in Idaho: southeastern part, mainly in Bonneville and Caribou counties
Land Use: woodland and watershed

The Prominent Characteristics of This Soil Are: E soil horizon at the surface; well developed subsoil

The first 3 soils in this book have had somewhat similar development. They all have light colored surface soils low in organic matter along with well-developed subsoils enriched with clay in climates which were moist. They differ mainly in having contrasting parent material. As will be seen, **Soil 2 (Santa Series)** has a special kind of subsoil.

Soil 1 has undergone extensive leaching in a cool moist climate. Little or no calcium carbonate remains of the original limestone parent material except in the lower part of the subsoil.

The surface layer lost clay and organic matter by leaching. It also became quite acid. A strongly acid soil limits the availability of plant nutrients for optimum growth of many plants. This is especially true when combined with the cool temperatures that exist during a short growing season. This contrasts with **Soil 3 (Porthill Series),** which has a much longer and warmer frost-free season, although it was leached in a similar manner.

Clay which was removed from the surface soil of **Soil 1** is now in the subsoil. This formed the light-colored E horizons comprised of mostly silt and sand particles. Clay films can easily be seen on soil aggregates in the Bt horizons with a hand lens. They resemble candle wax drippings.

The silty clay loam subsoil, with its clay films and prismatic and blocky structure, is well developed. It holds water well and root development is good although water movement is slow.

The use of this soil is mostly woodland with fair production. Some livestock graze on the understory. This soil is valuable as watershed for supplying much needed water for irrigation and power.

Crop production even on moderate slopes has little or no potential because of the cool very short growing season.

This soil is being used more and more for year-round recreation. The cool summers are a relief to those who live nearby in lower much warmer areas. This soil has good habitat for big game hunting in the fall. Snowmobiling and cross country skiing along logging roads has become more common.

Soil 2 (Santa Series)

A—0 to 5 inches; brown silt loam, granular structure, slightly acid.

Bw—5 to 24 inches; light brown silt loam, subangular blocky structure, slightly acid.

Ec—24 to 30 inches; very pale brown silt loam, prismatic structure, iron concretions, slightly acid.

Btxb1—30 to 37 inches; light brown silt loam, prismatic structure, prism faces coated with white silt coats and organic carbon, extremely hard and brittle, medium acid.

Btxb2—37 to 59 inches; light brown silt loam, prismatic and angular blocky structure, many dark brown clay films on prism and block faces, extremely hard and brittle, medium acid.

Soil 2 (Santa Series) with pasture in the foreground and wheat in the background. Soils formed in volcanic ash over sedimentary rock are on the hills in the distance.

Soil 2 (Santa Series)

Family Classification: coarse-silty, mixed, frigid Ochreptic Fragixeralfs

Soil Depth: 20 to 40 inches to fragipan

Drainage Class: moderately well drained

Parent Material: loess

Average Annual Precipitation: about 28 inches

Average Annual Air Temperature: about 43 degrees F.

Average Frost-Free Season: about 100 days

Elevation: about 2,800 feet

Topography: gently sloping to hilly uplands

Habitat Type: *Abies grandis/Pachystima myrsinites*

Common Native Vegetation: grand fir, Douglas-fir, ponderosa pine, lodgepole pine, western larch, western white pine, and myrtle pachystima

Occurrence in Idaho: northern part, mainly in Benewah, Kootenai, and Latah counties

Land Use: woodland and nonirrigated cropland

The Prominent Characteristics of This Soil Are: moderately deep to fragipan; E soil horizon below the surface soil; highly erodible surface soil; very slow permeability

This soil developed in loess. Parent material for this soil had its origin at least 200 miles away in Canada. During the ice age there were fingers of huge glaciers which slowly advanced into valleys of the United States. The melt water from these glaciers spread large amounts of silty alluvium over broad areas of south-central Washington. With no protective cover of vegetation the silt was easily picked up by strong winds during the dry summers and deposited onto the broad basaltic plateaus of eastern Washington and northern Idaho. Annual loess deposition still occurs, but with much less frequency and amount.

Other soils in this book which developed in loess of the same origin are **Soil 33 (Southwick Series), Soil 35 (Nez Perce Series),** and **Soil 50 (Palouse Series).**

Santa soils developed in the wettest coolest climates of these 4 soils. As a result, it differs in the kind of soil horizons it developed except for the E horizon below the surface. This is similar to the E horizon in the Southwick soils which formed at the junction of loess deposits of 2 different ages. The lower older loess having slower permeability than the new fresh loess caused water to accumulate at this point. With an excess of water, organic matter as well as clay was removed laterally leaving a leached light-colored layer. This layer becomes saturated each winter and spring and causes problems for crop production as well as causes basements to become wet.

The surface soil is low in organic matter and erodes easily when used for crop production unless extra control measures are taken.

The most striking feature of this soil is the subsoil. It is called a fragipan. It has high bulk density, a very low content of organic matter, and is seemingly cemented when dry. The high bulk density makes this soil about 1.8 times as heavy as an equal volume of water. During late summer and early fall digging by spade is nearly impossible and is even difficult with machinery. When moist it is somewhat brittle. It has very slow permeability. No roots penetrate natural soil aggregates — only a few are on vertical faces of large prisms.

Although the genesis of the fragipan is unknown, it could be due to the greater leaching which took place and the lack of freezing because of protection by deep snow cover. These two factors are in contrast to the other soils listed above. In any event its presence limits crop and timber production and complicates all other uses.

Soil 3 (Porthill Series)

E1—0 to 5 inches; light gray silt loam, granular structure, medium acid.

E2—5 to 15 inches; very pale brown silt loam, subangular blocky structure, medium acid.

Bt1—15 to 23 inches; pale olive silty clay loam, prismatic and angular blocky structure, medium acid upper part and moderately alkaline lower part.

Bt2—23 to 33 inches; pale yellow silty clay loam, platy structure, slightly effervescent, moderately alkaline.

Bk1—33 to 44 inches; pale olive silty clay loam, platy structure, strongly effervescent, moderately alkaline.

Bk2—44 to 59 inches; light olive gray silty clay loam, massive, strongly effervescent, moderately alkaline.

Soil 3 (Porthill Series) is in the foreground with hay and in the background with winter wheat and coniferous trees. The mountains in the distance were scoured by a huge, continental glacier.

Soil 3 (Porthill Series)

Family Classification: fine, mixed, frigid Typic Haploxeralfs

Soil Depth: 60 inches or more

Drainage Class: moderately well drained

Parent Material: glaciolacustrine

Average Annual Precipitation: about 21 inches

Average Annual Air Temperature: about 42 degrees F.

Average Frost-Free Season: about 125 days

Elevation: about 2,300 feet

Topography: nearly level and gently sloping high terraces

Habitat Type: *Thuja plicata/Pachystima myrsinites*

Common Native Vegetation: western redcedar, western larch, western white pine, Douglas-fir, myrtle pachystima, American trailplant, and starry false-Solomons-seal

Occurrence in Idaho: northern part, mainly in Boundary county

Land Use: woodland and nonirrigated cropland

The Prominent Characteristics of This Soil Are: E soil horizons at the surface; slow permeability

The Porthill soils developed in silty deposits which slowly settled out of a lake dammed by glacial ice. A lobe of the continental glacier which had pushed into northern Idaho from Canada was at this time in retreat. The Kootenai River was impounded for some time allowing for 300 to 500 feet of sediment to accumulate. After the ice further retreated the glacial lake quickly drained. The Kootenai River easily eroded through the soft silty deposits leaving a high nearly level terrace. This soil developed on this terrace.

The deposits came from a variety of sources, including limestone. The initial process of soil development was the leaching of lime from the upper parts of the soil. Further leaching took place removing clay from the surface soil and moving it into the subsoil. The subsoil now contains 35 to 50 percent clay.

The dense stands of conifers provided little organic matter to the soil. The surface soil remains light colored.

Most areas of this soil have been cleared of trees and are cultivated to wheat, oats, barley, hay, alfalfa seed, and Christmas trees. Production is high. However, since the organic matter level is naturally low, the crops need additional nitrogen to obtain these high yields. Most areas are nearly level. In places with sloping topography, this soil has a high erosion hazard.

Porthill soils also produce very good timber yields. The surface soil is soft when wet so equipment operations are restricted. Severe plant competition is a problem during reforestation.

The fine-textured subsoil has slow permeability which forms a perched water table. This affects the operation of septic tank filter fields for homesites.

Soil 4 (Chilcott Series)

E—0 to 5 inches; pale brown silt loam, platy structure, mildly alkaline.

Bt—5 to 26 inches; brown silty clay loam, subangular blocky structure, mildly alkaline.

Btk—26 to 32 inches; very pale brown silty clay loam, subangular blocky structure, strongly effervescent, mildly alkaline.

Bkqm—32 to 50 inches; white extremely hard duripan, strongly effervescent, moderately alkaline.

Bkq—50 to 59 inches; white weakly developed duripan, strongly effervescent, moderately alkaline.

Soil 4 (Chilcott Series) with Wyoming big sagebrush. Bennett mountains are in the background, with many rock outcrops.

Soil 4 (Chilcott Series)

Family Classification: fine, montmorillonitic, mesic Abruptic Xerollic Durargids

Soil Depth: 20 to 40 inches to duripan

Drainage Class: well drained

Parent Material: loess and alluvium

Average Annual Precipitation: about 9 inches

Average Annual Air Temperature: about 51 degrees F.

Average Frost-Free Season: about 150 days

Elevation: 2,500 to 3,500 feet

Topography: gently sloping plains

Habitat Type: *Artemisia tridentata wyomingensis/Agropyron spicatum*

Common Native Vegetation: Wyoming big sagebrush, bluebunch wheatgrass, Sandberg bluegrass, Thurber needlegrass, and arrowleaf balsamroot

Occurrence in Idaho: southwestern part, mainly from Gooding to Payette counties

Land Use: rangeland and irrigated cropland

The Prominent Characteristics of This Soil Are: light colored surface soil; well developed subsoil; strongly cemented duripan

The following 9 soils in this book have only 8 to 11 inches of precipitation, mostly falling in winter and spring months. Nonirrigated areas dry out early in the growing season so that normal crop production, even after a year of summer fallow, is generally impossible.

The Chilcott soils have well defined soil horizons, especially the lower lying lime-silica cemented duripan, which is indicative of a very old and stable land form. The arid area in which these soils occur has hot dry summers and cold winters. Weathering of minerals and the downward movement of clays, salts, carbonates, and soluble minerals by water in this environment has been very slow. The formation of the strongly-cemented duripan took place over many thousands of years. Slowly translocated silica and carbonates from the calcareous loess parent material formed the pan. Even though the climate at times during the development of this soil may have been more moist than the present, it was only wet enough to leach the silica and carbonates to the lower parts of the soil.

The presence of this pan has great significance to the use of this soil for cultivation. It limits root development and water penetration, except through a few cracks, and also limits the available water holding capacity. With the depth to the pan varying from 20 to 40 inches, this soil can hold only about 3½ to 7 inches of water that is available to plants. Also, the subsoil, which is slowly permeable, limits the rate at which water can be applied. Proper application of irrigation water is made difficult by these properties.

Irrigated crops include small grains, corn, sugar beets, alfalfa, clover, and improved pasture. Noncultivated areas used for rangeland have low forage production. The light-colored surface soil is low in organic matter. The natural vegetation of mostly bunchgrasses and Wyoming big sagebrush in this hot dry climate contributed little organic matter to the soil.

Soil 5 (Colthorp Series)

A—0 to 4 inches; pale brown very stony silt loam, platy and granular structure, mildly alkaline.

Bt—4 to 11 inches; light yellowish brown silt loam, subangular blocky structure, mildly alkaline.

Bk—11 to 16 inches; very pale brown very gravelly silty clay loam, subangular blocky structure, pieces of duripan, strongly effervescent, strongly alkaline.

Bkqm—16 to 24 inches; pinkish white extremely hard duripan, strongly effervescent, strongly alkaline.

2R—24 to 27 inches; basalt bedrock.

Soil 5 (Colthorp Series) with Wyoming big sagebrush. Lockman Butte is in the background.

Soil 5 (Colthorp Series)

Family Classification: loamy, mixed, mesic, shallow Xerollic Durargids

Soil Depth: 10 to 20 inches to duripan

Drainage Class: well drained

Parent Material: loess over basalt bedrock

Average Annual Precipitation: about 11 inches

Average Annual Air Temperature: about 51 degrees F.

Average Frost-Free Season: about 135 days

Elevation: 2,500 to 4,500 feet

Topography: gently sloping plains

Habitat Type: *Artemisia tridentata wyomingensis/Agropyron spicatum*

Common Native Vegetation: Wyoming big sagebrush, bluebunch wheatgrass, Sandberg bluegrass, Thurber needlegrass, arrowleaf balsamroot, and tapertip hawksbeard

Occurrence in Idaho: southwestern part, mainly in Ada and Elmore counties

Land Use: rangeland and irrigated cropland

The Prominent Characteristics of This Soil Are: shallow to duripan; moderately deep to bedrock

Colthorp soils are similar to **Soil 4 (Chilcott Series),** in parent material, climate, age, and sequence of development. They are, however, only moderately deep to basalt bedrock. The key similarity is the presence of a lime-silica cemented duripan which, like the pan in the Chilcott soils, took many thousands of years to develop. **Soil 39** also has a duripan but it was cemented only by silica.

The pan in the Colthorp soils is thinner than the pan in the Chilcott soils but it is just as effective in preventing the downward movement of water and the downward extension of roots. It occurs at a more shallow depth and thus the soil holds less total available water. The depth to the duripan ranges from only 10 to 20 inches so the soils can hold only about 1½ to 3½ inches of available water. This severely limits the suitability of these soils for irrigation. Cultivating these soils is justified only where it occurs with deeper more desirable soils. Crop yields of small grain, corn, sugar beets, potatoes, hay, and pasture are low.

Greater use is made of these soils for rangeland, however forage production is low. Because of the low production these soils are susceptible to overgrazing which decreases the grasses and increases the growth of the sagebrush.

The presence of basalt bedrock at depths ranging from 20 to 40 inches affects many engineering uses of these soils. Whenever excavations must be made for basements, pipelines, and the like, the expense is great. Avoid these soils for these uses if possible.

Soil 6 (Sebree Series)

E—0 to 6 inches; very pale brown silt loam, platy and granular structure, slightly acid.

Btn—6 to 19 inches; light brown clay, prismatic and angular blocky structure, high in sodium, neutral.

Bkz—19 to 24 inches; pink clay loam, prismatic structure, violently effervescent, moderately alkaline.

Bkqm—24 to 32 inches pink extremely hard duripan, violently effervescent, moderately alkaline.

2Bkq—32 to 44 inches; light yellowish brown loamy sand, platy structure, strongly effervescent, moderately alkaline.

2Btb—44 to 60 inches; pale brown loamy sand, subangular blocky structure, mildly alkaline.

Soil 6 (Sebree Series) showing poor growth of corn in small area near the center of photograph. Soils surrounding this small area are not high in sodium and corn growth is good.

Soil 6 (Sebree Series)*

Family Classification: fine-silty, mixed, mesic Xerollic Nadurargids

Soil Depth: 20 to 40 inches to duripan

Drainage Class: well drained

Parent Material: loess and alluvium

Average Annual Precipitation: about 10 inches

Average Annual Air Temperature: about 47 degrees F.

Average Frost-Free Season: about 140 days

Elevation: 2,500 to 4,500 feet

Topography: nearly level to moderately sloping plains

Habitat Type: barren areas within *Artemisia tridentata wyomingensis/Agropyron spicatum*

Common Native Vegetation: barren or nearly so; some stunted Wyoming big sagebrush, cheatgrass, and pepperweed are around the edges of these soils

Occurrence in Idaho: southwestern part, mainly from Elmore to Payette counties

Land Use: barren areas within associated soils that are in rangeland and irrigated cropland

The Prominent Characteristics of This Soil Are: barren soil under natural conditions; E soil horizon at the surface; sodium in the subsoil; strongly cemented duripan

The Sebree Series occurs in small areas only a few feet across. Where uncultivated these areas are void or nearly so of any natural vegetation. They appear quite light colored on the surface. The subsoil, which lies close to the surface, is high in sodium. This causes the clay to disperse, resulting in very poor tilth. Permeability of the subsoil is very slow so moisture barely enters the soil. Areas of this soil are referred to as "slick spots" because the surface seems to puddle or "slicken-over" with little precipitation.

This soil underwent a complex development involving soil chemistry and water movement. It had impeded drainage and formed in a rather warm arid climate. The natural weathering of the soil released soluble salts, which in the dry climate rose by capillary action rather than leaching downward as is normal with well drained soils. Water movement carried soluble salts from adjacent soils into the slick spots. More and more sulfates and chlorides of sodium, calcium, magnesium, and potassium were brought toward the surface, with sodium salts being most prominent. Sodium caused the clay to act as individual particles rather than clinging together in soil aggregates which is normal.

Very little can be done to the Sebree soil to improve its condition if left uncultivated. Where cultivated this soil can be improved at least to some extent by incorporating large amounts of organic materials and gypsum. This improves the physical condition allowing water to pass through the soil. If the soil is irrigated as well, this will speed the leaching of the offending sodium salts out of the soil.

Another method sometimes used to improve this soil, however costly, is deep plowing up to 30 inches deep. This turns the layers, which are high in sodium, on edge which allows water to penetrate.

The duripan developed over many thousands of years, long before the present soil formed. It is similar to the duripan in **Soil 4 (Chilcott Series).**

Soils similar to Sebree Series occur near Lewiston in northern Idaho, but they are all dryfarmed. Their improvement will take longer without irrigation to help leach the sodium.

*This soil is outside the defined limits of the Sebree Series because the E horizon is thicker than 2 inches.

Soil 7 (Gooding Series)

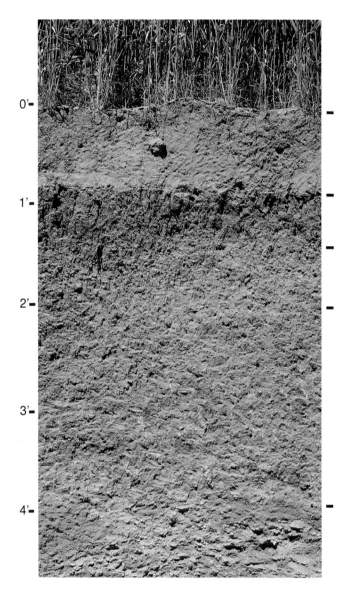

Ap—0 to 10 inches; pale brown silt loam, granular structure, few gravel, mildly alkaline.

2Bt1—10 to 16 inches; brown silty clay, prismatic and angular blocky structure, mildly alkaline.

2Bt2—16 to 23 inches; light brown silty clay loam, angular blocky structure, moderately alkaline.

2Btk—23 to 46 inches; light brown silty clay loam, subangular blocky structure, slightly effervescent, moderately alkaline.

3Bk—46 to 56 inches; pinkish white sandy loam, massive, partially wetted by irrigation, violently effervescent, strongly alkaline.

Soil 7 (Gooding Series) with winter wheat.

Soil 7 (Gooding Series)

Family Classification: fine, montmorillonitic, mesic Xerollic Paleargids
Soil Depth: 40 to 60 inches or more to basalt bedrock
Drainage Class: well drained
Parent Material: alluvium over loess over basalt residuum
Average Annual Precipitation: about 9 inches
Average Annual Air Temperature: about 49 degrees F.
Average Frost-Free Season: about 125 days
Elevation: 2,500 to 4,500 feet
Topography: nearly level to rolling plains
Habitat Type: *Artemisia tridentata wyomingensis/Agropyron spicatum*
Common Native Vegetation: Wyoming big sagebrush, bluebunch wheatgrass, and Sandberg bluegrass
Occurrence in Idaho: southern part, mainly in Gooding county
Land Use: rangeland and irrigated cropland

The Prominent Characteristics of This Soil Are: light colored surface soil; well developed subsoil; very slow permeability; lime accumulation deep in the soil

Gooding soils exhibit a very complex soil development which is not apparent from a view of the surface. Three parent materials are involved — recent alluvium over loess over basalt residuum. Little development of the basalt residuum took place. The loess, when it first blanketed the basalt residuum, was calcareous. Lime was leached to moderate depths because of the relatively dry climate. This was quickly followed by a long period of weathering and a downward movement of newly formed clay. This slowly increased the percentage of clay in the B horizons from about 10 or 15 percent to more than 40 percent, especially in the upper parts.

Then during an ice advance into northern Idaho from Canada, at which time the climate in southern Idaho was also much colder and wetter, these soils went through a process similar to that described in **Soil 14 (Flybow Series).** The original A horizon and the upper part of the B horizon was eroded away during this cold wet period. Several inches of fresh silty alluvium then washed upon these soils. In the photograph, a piece of gravel in the surface soil is the clue that shows this to be alluvium.

A thin E horizon then formed in the lower part of the alluvium. This was because the permeability of the old clayey B horizon was much less than that of the overlying alluvium. A temporary perched water table at this point leached clay and organic matter and formed the E horizon. Cultivation has obliterated most of this horizon, however, a remnant of it can be seen on the left side in the photo.

The light colored surface soil is low in organic matter. Dry climates produce only sparse vegetation which when decomposed forms little organic matter.

Where these soils are used for rangeland the forage production is moderately good if the range is in good condition. Reseeding areas in poor condition is possible but success may be low due to the shallow depth to the claypan.

When these soils are cultivated, irrigation is necessary because of the dry climate. Crops include corn, sugar beets, wheat, and alfalfa hay. Yields are good.

Many homesites have been developed on these soils. The very slow permeability of the subsoils causes problems with septic systems.

Soil 8 (Portneuf Series)

A—0 to 5 inches; pale brown silt loam, platy structure, strongly effervescent, moderately alkaline.

Bw—5 to 16 inches; light yellowish brown silt loam, prismatic structure, strongly effervescent, moderately alkaline.

Bk—16 to 30 inches; white silt loam, massive, violently effervescent, moderately alkaline.

Bkqz—30 to 40 inches; very pale brown silt loam, prismatic structure, many silica cemented nodules, violently effervescent, moderately alkaline.

C—40 to 63 inches; very pale brown silt loam, massive, slightly effervescent, moderately alkaline.

Soil 8 (Portneuf Series) with irrigated alfalfa.

Soil 8 (Portneuf Series)

Family Classification: coarse-silty, mixed, mesic Durixerollic Calciorthids
Soil Depth: 40 to 60 inches or more to basalt bedrock
Drainage Class: well drained
Parent Material: loess
Average Annual Precipitation: about 9 inches
Average Annual Air Temperature: about 48 degrees F.
Average Frost-Free Season: about 130 days
Elevation: 2,500 to 4,700 feet
Topography: gently and moderately sloping plains
Habitat Type: *Artemisia tridentata wyomingensis/Stipa thurberiana*
Common Native Vegetation: Wyoming big sagebrush, rabbitbrush, bluebunch wheatgrass, Sandberg bluegrass, Thurber needlegrass, and prairie junegrass
Occurrence in Idaho: southern part, mainly from Twin Falls to Bingham counties
Land Use: rangeland and irrigated cropland

The Prominent Characteristics of This Soil Are: light colored surface soil; lime accumulation below the surface soil

Portneuf soils are very important agricultural soils in southern Idaho. They occur, however, in a climate which is too dry to be farmed without irrigation. Temperatures are warm enough, though, for a wide variety of crops. These include dry beans, sugar beets, potatoes, corn, alfalfa hay, wheat, and barley, with high yields.

The soils are deep and very deep and have high total available water capacities. This, along with good infiltration rates, makes for easy irrigation management. The light-colored surface soil is low in organic matter so fertilizers are necessary for good crop yields. Zinc deficiencies are common, especially for beans in places where the soils have been leveled for surface irrigation. This exposes the more limey subsurface horizons. Salts also occur deep in the soils. This could be a problem in the future with drainage waters becoming more salty and affecting soils and plants downstream.

A few areas are used for rangeland. Many places have been overgrazed and are now in poor range condition. These soils can be reseeded for improvement.

Portneuf soils developed mostly in loess, but in some areas they have been developed in similar textured lacustrine material that has been reworked by wind. Low precipitation in the dry climate, prevailing on these soils, leached the lime present in the parent material only to the usual depth of water penetration. In places there is still some lime remaining in the surface soil. The surface soil, like many soils with a similar dry climate such as **Soil 13 (Garbutt Series),** is light colored. This is because the sparse natural vegetation provided little organic matter to darken the soil.

Rounded nodules of very hard soil material make up 20 to 60 percent of parts of the Bkqz horizon. They are weakly cemented by silica. This represents soil development that, given enough time, will develop a duripan such as that in **Soil 11 (Minidoka Series).**

Soil 9 (Trevino Series)

A—0 to 3 inches; brown very fine sandy loam, granular structure, moderately alkaline.

Bw—3 to 10 inches; pale brown silt loam, subangular blocky structure, moderately alkaline.

Bk—10 to 17 inches; very pale brown cobbly silt loam, massive, strongly effervescent, moderately alkaline.

2R—17 to 30 inches; basalt bedrock.

Soil 9 (Trevino Series) shown in road cut.

Soil 9 (Trevino Series)

Family Classification: loamy, mixed, mesic Lithic Xerollic Camborthids
Soil Depth: 10 to 20 inches to basalt bedrock
Drainage Class: well drained
Parent Material: loess mixed with basalt
Average Annual Precipitation: about 9 inches
Average Annual Air Temperature: about 47 degrees F.
Average Frost-Free Season: about 135 days
Elevation: 2,000 to 4,000 feet
Topography: gently and moderately sloping plains
Habitat Type: *Artemisia tridentata wyomingensis/Agropyron spicatum*
Common Native Vegetation: Wyoming big sagebrush, bluebunch wheatgrass, Sandberg bluegrass, Thurber needlegrass, arrowleaf balsamroot, and tapertip hawksbeard
Occurrence in Idaho: southern part, mainly from Canyon to Minidoka counties
Land Use: rangeland and irrigated cropland

The Prominent Characteristics of This Soil Are: shallow to bedrock; low available water holding capacity

This soil is similar to **Soil 43 (Gwin Series)** and **Soil 48 (Hymas Series).** All of these soils are shallow to bedrock. Gwin soils have well developed Bt horizons and Hymas soils developed from limestone and are calcareous throughout. The shallowness to bedrock is the overriding feature of all of these soils which greatly restricts their use and productivity.

Excavations for basements, pipelines, road building and the like are made quite difficult and therefore expensive. In planning for these kinds of uses, it is wise to examine the soil at the site for this feature. It may be possible to move a few feet to avoid such a problem.

Bedrock at this shallow depth not only limits root development below a depth of 10 to 20 inches to a few cracks in the bedrock, but it greatly reduces the total available water capacity, thus affecting yields. Where Trevino soils are cultivated, irrigation is required because of a lack of sufficient rainfall. Since these soils are commonly cultivated with nearby deep soils with high water holding capacities, the design of the irrigation system tends to ignore the shallow soils and so the crops suffer.

Where these soils are used for range, the forage production is low. The range is most often in poor condition because of overgrazing.

Trevino soils developed in a dry climate similar to **Soil 8 (Portneuf Series).** Both soils have light colored surface soils low in organic matter and both soils have lime accumulations below the surface.

Soil 10 (Owyhee Series)

A—0 to 9 inches; light brownish gray silt loam, granular structure, moderately alkaline.

Bw—9 to 19 inches; pale brown silt loam, blocky structure, slightly effervescent; moderately alkaline.

Bk—19 to 31 inches; light gray silt loam, blocky structure, violently effervescent, moderately alkaline.

C—31 to 53 inches; light gray silt loam, platy lacustrine, slightly effervescent, moderately alkaline.

Soil 10 (Owyhee Series) with a crop of irrigated onions. Soils on the hills in the background are shallow to deep to bedrock and contain many rock fragments.

Soil 10 (Owyhee Series)

Family Classification: coarse-silty, mixed, mesic Xerollic Camborthids

Soil Depth: 60 inches or more

Drainage Class: well drained

Parent Material: lacustrine

Average Annual Precipitation: about 9 inches

Average Annual Air Temperature: about 52 degrees F.

Average Frost-Free Season: about 150 days

Elevation: 2,200 to 2,800 feet

Topography: nearly level and gently sloping terraces

Habitat Type: *Artemisia tridentata wyomingensis/Agropyron spicatum*

Common Native Vegetation: Wyoming big sagebrush, rabbitbrush, bluebunch wheatgrass, Sandberg bluegrass, and Thurber needlegrass

Occurrence in Idaho: southwestern part, mainly in Canyon, Elmore, Payette, and Washington counties

Land Use: irrigated cropland

The Prominent Characteristics of This Soil Are: slow permeability; laminated substratum; leached upper subsoil; lime accumulation deep in the soil

The Owyhee soils developed in silty material which slowly accumulated in lakes. At a later date, the impounded waters drained completely and exposed a new lacustrine parent material to be modified by the action of climate and living organisms. The present soil resulted. It has a similar parent material as that of **Soil 3 (Porthill Series),** except that the lake deposits of the Porthill soils came from a lake formed by the damming of a valley by glacial ice. In either case there was an accumulation of silty deposits.

The platy or laminated lacustrine layers are less prominent toward the surface because of the effects of soil forming processes after the sediments were exposed. Mixing by animal activity, root development, frost action, and the reworking by wind and water all tend to destroy the original platiness of the parent material.

With an arid climate, the calcareous deposits were only partially leached. The surface soil and upper subsoil remains moderately alkaline and an accumulation of calcium carbonate occurs just above the most prominent laminations. The light-colored soil horizon of enriched lime may look like an E soil horizon which results from intense leaching, such as in **Soil 33 (Southwick Series).** But a drop of dilute acid soon tells the difference. A soil horizon with calcium carbonate thus treated will effervesce, giving off carbon dioxide gas in bubbles.

The Owyhee soil shows the very beginnings of a well-developed subsoil just below the surface layer. The Bw soil horizon has lost most of its lime and has gained some brown color along with a blocky structure. Given enough time, it will lose all of its lime and accumulate clay, like **Soil 7 (Gooding Series).**

Most areas of the Owyhee series are being irrigated to produce a wide variety of crops. With an ideal climate of warm summer night temperatures and a good growing season, crops include wheat, potatoes, corn, sugar beets, mint, onions, vegetable seed crops, orchards, hay, and pasture. Yields are excellent.

Slow permeability within the laminated material affects some uses of these soils such as septic tank absorption fields.

Soil 11 (Minidoka Series)

A—0 to 7 inches; very pale brown silt loam, blocky structure, slightly effervescent, moderately alkaline.

Bk1—7 to 20 inches; very pale brown loam, blocky structure, strongly effervescent, moderately alkaline.

Bk2—17 to 36 inches; white loam, massive, violently effervescent, moderately alkaline.

Bkqmb—36 to 55 inches; extremely hard fractured duripan, moderately alkaline.

Soil 11 (Minidoka Series) with irrigated sugar beets.

Soil 11 (Minidoka Series)

Family Classification: coarse-silty, mixed, mesic Xerollic Durorthids

Soil Depth: 20 to 40 inches to duripan

Drainage Class: well drained

Parent Material: loess or silty alluvium

Average Annual Precipitation: about 9 inches

Average Annual Air Temperature: about 50 degrees F.

Average Frost-Free Season: about 135 days

Elevation: 2,000 to 5,000 feet

Topography: nearly level and gently sloping terraces or plains

Habitat Type: *Artemisia tridentata wyomingensis/Stipa thurberiana*

Common Native Vegetation: Wyoming big sagebrush, rabbitbrush, bluebunch wheatgrass, Thurber needlegrass, Sandberg bluegrass, lupine, and arrowleaf balsamroot

Occurrence in Idaho: southern part, mainly from Canyon to Minidoka counties

Land Use: rangeland and irrigated cropland

The Prominent Characteristics of This Soil Are: strongly cemented duripan; light colored surface soil

The Minidoka soils have a strongly cemented duripan which is similar to the pan in **Soil 4 (Chilcott Series).** Like all thick lime-silica cemented pans it developed over many thousands of years. There was only enough moisture in this arid climate to leach the carbonates and silica to lower depths. The cementing action of these agents slowly formed the pan. The pan thickened as this process continued.

This duripan affects most uses of this soil. It limits the available moisture capacity to the soil above the pan. With only 20 to 40 inches of soil above the pan, there is only 3½ to 7 inches of available water capacity. Forage production on the natural vegetation, even if in good condition, is restricted. When these soils are used for field crops the proper application of irrigation water is made more difficult by this lower available water capacity.

The normal rainfall is too low for cultivation without irrigation. The main crops on these soils are dry beans, sugar beets, potatoes, corn, and wheat. A small acreage is used for hay and pasture. The natural fertility is low but production is good with the use of fertilizer and the proper application of water. Cut areas from land leveling causes zinc deficiencies in some crops.

The parent material of the Minidoka soils above the pan has been reworked by wind and water. This part of the soil is young in contrast to the pan which is very old. The surface soil remains somewhat calcareous.

Like the Chilcott soils, the light-colored surface layer is low in organic matter. The hot dry climate produced a sparse vegetation of mostly Wyoming big sagebrush and bunchgrasses. This contributed little organic matter unlike the cooler more moist climates of the prairies which had a rank growth of grasses, as in **Soil 50 (Palouse Series).**

Irrigated farms on the Minidoka soils are small so there are many houses. The thick cemented pan is a problem when making shallow excavations and in developing successful sewage facilities.

Soil 12

A—0 to 7 inches; pale brown gravelly silt loam, granular structure, slightly effervescent, moderately alkaline.

Bk—7 to 15 inches; pale brown gravelly silt loam, subangular blocky structure, strongly effervescent, moderately alkaline.

Bkqm1—15 to 21 inches; white extremely hard duripan, strongly effervescent, moderately alkaline.

Bkqm2—21 to 37 inches; very pale brown duripan, strongly effervescent, moderately alkaline.

Bkq—37 to 62 inches; multi-colored extremely gravelly loamy coarse sand, strongly effervescent, moderately alkaline.

Soil 12 is in the foreground on broad fan terraces at the base of limestone mountains.

Soil 12

Family Classification: loamy, mixed, shallow, frigid * Xerollic Durorthids
Soil Depth: 10 to 20 inches to duripan
Drainage Class: well drained
Parent Material: alluvium
Average Annual Precipitation: about 10 inches
Average Annual Air Temperature: about 41 degrees F.
Average Frost-Free Season: About 80 days
Elevation: 6,000 to 7,000 feet
Topography: gently and moderately sloping fan terraces
Habitat Type: *Artemisia arbuscula/Agropyron spicatum*
Common Native Vegetation: low sagebrush, bluebunch wheatgrass, and Sandberg bluegrass
Occurrence in Idaho: eastern part, mainly in Butte and Clark counties
Land Use: rangeland

The Prominent Characteristics of This Soil Are: shallow to duripan; duripan formed in coarse fragments; light-colored surface soil

This soil occurs along the sides of broad mountain valleys on gently and moderately sloping fan terraces. These terraces have the same shape as those in which **Soil 45 (Little Wood Series)** developed. Both soils are about alike in having an abundance of somewhat rounded gravel and cobblestones which were washed out of nearby mountains during a much wetter period than the present. Little Wood soils developed in an alluvial mixture of rock material whereas the alluvium of **Soil 12** came from limestone mountains.

The climate became drier and the flooding ceased, leaving a landscape which has changed very little over a long period of time. These two soils developed into contrasting soils because of the different effects of climate and parent material.

The parent material of the Little Wood soils contained little calcium carbonate. What little there was quickly leached out of the soil. The limestone material of **Soil 12** provided a large supply of calcium carbonate. Lime is still present throughout the soil.

The average annual precipitation is now about 14 inches on the Little Wood soils and about 10 inches on **Soil 12**. This may not seem like much of a difference but when moisture is limited, it can mean a lot, especially if the soils contain many coarse fragments. The more abundant native vegetation on the Little Wood soils, produced by the higher precipitation, provided more organic matter with which to darken the soil.

The arid climate of **Soil 12** developed a lime-silica cemented duripan at a shallow depth similar to that in **Soil 5 (Colthorp Series)**. This limited the vegetative growth even more, so that the surface soil is low in organic matter and light colored. Soil moisture may be high at the time of snow melt in the spring but the soil dries out rapidly and thoroughly early in the growing season. It is important to realize that if any desirable grazing species are eliminated by overgrazing, it would be very difficult to re-establish them in this harsh environment.

*This soil has cryic temperatures that are not presently considered in classification (1).

Soil 13 (Garbutt Series)

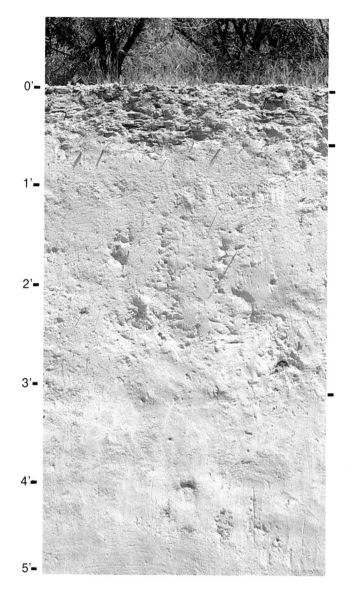

A—0 to 6 inches; pale brown silt loam, platy structure, slightly effervescent, moderately alkaline.

C1—6 to 36 inches; pale brown silt loam, massive, strongly effervescent, strongly alkaline.

C2—36 to 60 inches; pale brown silt loam, massive, strongly effervescent, very strongly alkaline.

Soil 13 (Garbutt Series) being irrigated for wheat. Cache Peak is in the distance.

Soil 13 (Garbutt Series)

Family Classification: coarse-silty, mixed (calcareous), mesic Typic Torriorthents

Soil Depth: 60 inches or more

Drainage Class: well drained

Parent Material: alluvium

Average Annual Precipitation: about 8 inches

Average Annual Air Temperature: about 51 degrees F.

Average Frost-Free Season: about 135 days

Elevation: 2,000 to 5,400 feet

Topography: nearly level and gently sloping fans and low terraces

Habitat Type: *Eurotia lanata/Oryzopsis hymenoides*

Common Native Vegetation: winterfat, shadscale, Indian ricegrass, bottlebrush squirreltail, Sandberg bluegrass, and Thurber needlegrass

Occurrence in Idaho: southern part, mainly from Canyon to Minidoka counties

Land Use: rangeland and irrigated cropland

The Prominent Characteristics of This Soil Are: light-colored surface soil; no subsoil development

The Garbutt soils developed in fairly recent silty alluvium derived mostly from old calcareous lacustrine siltstone. These soils, along with **Soil 14 (Flybow Series), Soil 15 (Pyle Series), Soil 16 (Quincy Series),** and **Soil 17 (Shellrock Series),** represent the least soil development that has taken place since the parent material was put into place. This is due to the recent nature of the parent material. It not only takes time for soils to weather and form B horizons, but also to leach lime and salts present in the parent material and to accumulate organic matter in the surface soils. The hot dry climate of the Garbutt soils provided little excess moisture to leach lime. Note that the surface soil still has lime present. This climate also supported only sparse natural vegetation which provided little organic matter to the soils.

The light-colored surface soil, low in organic matter, is typical of many soils in southern Idaho with an arid climate. These include soils such as **Soil 5 (Colthorp Series)** and **Soil 8 (Portneuf Series).**

Most areas of **Soil 13 (Garbutt Series)** are under irrigation. Crops include sugar beets, potatoes, corn, wheat, alfalfa hay, and pasture. Yields are high. The total available water capacity is very high, no layers in the soils impede water movement and root development, and the climate provides for optimum growth. The soil is easy to cultivate and with properly designed irrigation systems the erosion hazard is minimal. The only potential problem that may arise in the future is from the salt present in the subsurface horizons.

When this soil is used for rangeland, the usable forage production is low. Cheatgrass easily invades areas poorly managed, however the soils can easily be reseeded.

The soil is suitable for most building site development.

Soil 14 (Flybow Series)

A—0 to 4 inches; brown very cobbly loam, subangular blocky and granular structure, slightly acid.

R—4 to 34 inches; basalt bedrock.

Soil 14 (Flybow Series) is in the foreground and in the intermound area. The mounds consist of moderately deep and deep loamy soils. This kind of landscape is called "patterned ground".

Soil 14 (Flybow Series)

Family Classification: loamy-skeletal, mixed, nonacid, mesic Lithic Xerorthents

Soil Depth: 2 to 10 inches to basalt bedrock

Drainage Class: well drained

Parent Material: basalt residuum

Average Annual Precipitation: about 19 inches

Average Annual Air Temperature: about 48 degrees F.

Average Frost-Free Season: about 150 days

Elevation: 1,000 to 4,400 feet

Topography: moderately sloping to steep south-facing canyons

Habitat Type: *Agropyron spicatum/Opuntia polyacantha*

Common Native Vegetation: bluebunch wheatgrass, Idaho fescue, Sandberg bluegrass, plains pricklypear, cutleaf balsamroot, biscuitroot, and lupine

Occurrence in Idaho: north-central part, mainly from Latah to Idaho counties

Land Use: rangeland

The Prominent Characteristics of This Soil Are: very shallow to bedrock; intermound soil

The Flybow soil is one of the few soils in Idaho with very limited uses. The overriding feature of having less than 10 inches of very cobbly loam soil overlying basalt bedrock restricts its use to rangeland. Its annual forage production is very low. Some use is made of the bedrock underlying this soil as sites for rock quarries. All other uses to be made of this soil have very severe limitations.

Here is an excellent example of the necessity for careful examination of a soil to determine its potential for use. Many other soils may look like this one when only seeing the surface. It might seem that a simple job of removing the surface cobblestones would be all that is necessary. However, the presence of bedrock at a very shallow depth is hidden from view.

The ice age is responsible for the present character of this soil. Loess coming from areas of glacial outwash some distance away attained depths of about 3 feet overlying basalt on these south-facing positions. The cold climate which prevailed during the ice age kept the saturated loess frozen most of the time with periodic thawing during summer months. This caused uneven erosion, which formed a landscape of mounds and intermounds called "patterned ground". The very shallow Flybow soil is in the intermound areas which were eroded leaving the frozen soil as mounds of somewhat deeper soils. The mounds are 50 to 100 feet in diameter with the intermound areas 50 to 150 feet wide.

As the slopes become steeper, the mounds become elongated down the slope. On some steep slopes, Flybow soil occurs with few mounds of deeper soil remaining.

Mound and intermound topography occurs around the world near the limits of ice advance on all kinds of parent material. This process is active today in the far north.

Soil 15 (Pyle Series)

A—0 to 6 inches; grayish brown loamy coarse sand, granular structure, neutral.

Bw—6 to 20 inches; pale brown loamy coarse sand, subangular blocky structure, slightly acid.

C—20 to 27 inches; very pale brown gravelly sand, single grain, few lamellae, slightly acid.

Cr—27 to 52 inches; unconsolidated granodiorite bedrock.

Soil 15 (Pyle Series) is on the Douglas-fir covered steep slopes in the background. The nearly level area in the foreground has poorly drained soils with a high water table.

Soil 15 (Pyle Series)

Family Classification: mixed Alfic Cryopsamments

Soil Depth: 20 to 40 inches to granodiorite bedrock

Drainage Class: somewhat excessively drained

Parent Material: granodiorite, quartz monzonite, or quartz diorite residuum

Average Annual Precipitation: about 30 inches

Average Annual Air Temperature: about 40 degrees F.

Average Frost-Free Season: 50 days or less

Elevation: 5,000 to 7,000 feet

Topography: steep and very steep foothills and mountains

Habitat Type: *Pseudotsuga menziesii/Physocarpus malvaceus*

Common Native Vegetation: Douglas-fir, ponderosa pine, mallow ninebark, Saskatoon serviceberry, pine reedgrass, sedge, and longtube twinflower

Occurrence in Idaho: central part, mainly in Boise and Valley counties

Land Use: woodland and watershed

The Prominent Characteristics of This Soil Are: very low available water holding capacity; easily rippable bedrock; sandy textures

The Pyle soils are weakly developed. A thin A soil horizon is present with only a moderate amount of organic matter. The sandy Bw horizon has had little weathering to form clay.

Weathering of granite-like bedrock, such as under the Pyle soils, is slow in a cool moist climate. It has weathered, however, to the point that it is rippable by equipment. Other soils having rippable bedrock are **Soil 17 (Shellrock Series), Soil 27,** and **Soil 49 (Ola Series).** They all developed from granite or granite-like bedrock. Road building is much easier in such materials when they are weathered in comparison to the hard basalt bedrock underlying soils such as **Soil 42 (Gem Series)** and **Soil 52 (Magic Series).** Much heavier equipment or blasting would be needed on the Gem and Magic soils.

Soil textures above the apparent bedrock are sandy from the granitic parent material. Clay content is low. Due to the sandy textures and limited soil depth, the total available water holding capacity is very low — less than 3 inches. Production of the woodland is limited by this droughtiness but is partially overcome by the moderately high precipitation and low evapotranspiration.

Some woodland grazing is possible, however the droughtiness limits forage production. Also, the habitat for woodland wildlife is only fair.

Pyle soils developed in parent material similar to **Soil 17 (Shellrock Series).** Pyle soils, having a cooler more moist climate, have a lighter-colored surface layer. Understory plants, which provide most of the organic matter in the soils, were less abundant. This difference in surface soil colors is common with soils developed in similar parent materials, but having different climates.

Soil 16 (Quincy Series)

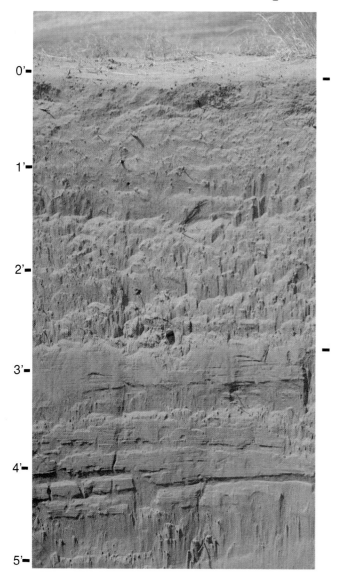

C1—0 to 34 inches; light gray fine sand, single grain, loose, moderately alkaline.

C2—34 to 60 inches; light gray fine sand, single grain, loose, lenses of very fine sand which retain more moisture, moderately alkaline.

Soil 16 (Quincy Series) is in foreground. Large active sand dune is in the background.

Soil 16 (Quincy Series)

Family Classification: mixed, mesic Xeric Torripsamments

Soil Depth: 60 inches or more

Drainage Class: excessively drained

Parent Material: eolian sand

Average Annual Precipitation: about 8 inches

Average Annual Air Temperature: about 52 degrees F.

Average Frost-Free Season: about 150 days

Elevation: about 2,500 feet

Topography: gently sloping to steep hummocks or dunes

Habitat Type: *Artemisia tridentata tridentata/Oryzopsis hymenoides*

Common Native Vegetation: basin big sagebrush, antelope bitterbrush, Indian ricegrass, needleandthread, thickspike wheatgrass, Sandberg bluegrass, and bluebunch wheatgrass

Occurrence in Idaho: southern part, mainly from Ada to Bingham counties

Land Use: rangeland and irrigated cropland

The Prominent Characteristics of This Soil Are: sandy textures; no soil development; low available water holding capacity; rapid permeability

This soil is the least-developed soil discussed in this book. No B horizons developed, not even an A horizon. The sands of this dune topography are slowly stabilized by plants. This soil qualifies as being "soil" since it now supports plants. The more active dune shown in the distance in the landscape photo is not "soil".

The soil profile photo shows how layers of slightly different sizes of sand were laid down by the wind. Very little clay exists to help bind the soil particles together. The soil has no structure and is loose, similar to a sandy beach. Permeability is very rapid and the total available water capacity is low. There is almost no runoff of water because the soil is so porous. Wind erosion is a serious problem when the soil is dry and vegetation is removed.

Most areas of this soil are used as rangeland, left idle, or used as wildlife habitat. The usable forage is not only low in amount, but is very susceptible to destruction if grazed.

In some areas irrigated crops are grown such as corn, potatoes, wheat, and alfalfa hay. Careful irrigation water management is needed because of the low available water and nutrient holding capacity. Frequent applications of small amounts of water and fertilizer are needed. This soil is especially suited to center pivot irrigation systems.

With high percolation rates, groundwater pollution occurs when building sites use improperly designed septic systems. The droughtiness of this soil makes establishment of lawns and other plants difficult.

Soil 17 (Shellrock Series)

A—0 to 6 inches; dark grayish brown loamy coarse sand, granular structure, soft, medium acid.

C1—6 to 19 inches; brown loamy coarse sand, single grain, loose, medium acid.

C2—19 to 37 inches; yellowish brown coarse sand, single grain, slightly acid.

Cr—37 to 62 inches; weathered granite.

Soil 17 (Shellrock Series) with ponderosa pine. Granite boulders are common.

Soil 17 (Shellrock Series)

Family Classification: mixed, frigid Typic Xeropsamments

Soil Depth: 40 to 60 inches to slightly weathered granite

Drainage Class: somewhat excessively drained

Parent Material: granite residuum

Average Annual Precipitation: about 27 inches

Average Annual Air Temperature: about 41 degrees F.

Average Frost-Free Season: about 80 days

Elevation: 4,500 to 6,500 feet

Topography: moderately sloping to steep ridges and south-facing foothills and mountains

Habitat Type: *Pinus ponderosa/Purchia tridentata, Festuca idahoensis phase*

Common Native Vegetation: ponderosa pine, antelope bitterbrush, snowberry, snowbrush ceanothus, Idaho fescue, and bluebunch wheatgrass

Occurrence in Idaho: central part, mainly in Valley county

Land Use: woodland and watershed

The Prominent Characteristics of This Soil Are: no subsoil development; sandy textures; easily rippable bedrock; very low available water holding capacity

Shellrock soils developed in residuum from granite. Little development has taken place other than forming a thin A horizon. No underlying B horizons, which occur in more developed soils, are present. Cool temperatures have kept weathering at a very slow rate. At the same time the natural geologic erosion of the steep slopes kept removing soil material at about the same rate at which weathering took place. This has kept the soils young.

This soil developed in parent material similar to **Soil 15 (Pyle Series).** Shellrock soils have less precipitation and are a bit warmer in summer months than the Pyle soils. Stands of ponderosa pine on Shellrock soils are more open than the mixture of Douglas-fir and ponderosa pine on the Pyle soils. This allows more understory plants, especially grasses, to grow and results in a thicker and darker colored surface soil containing more organic matter.

Shellrock soils, forming from granite, are sandy throughout. The low amount of clay and silt makes these soils very droughty. The total available water capacity is less than 3 inches. The growth of trees, as well as the grazable forage, is limited by this natural droughtiness. This also affects wildlife habitat because of limited feed and cover.

These soils are used mostly for woodland. Small seedlings have difficulty becoming established after timber harvest, due to the droughtiness of the soils. Some areas of woodland are grazed by livestock but most are left for wildlife, recreation, and watershed. All other uses are greatly limited by the steep slopes.

Soil 18 (Pywell Series)

Op—0 to 7 inches; black sapric material, no fibers present after rubbing, granular structure, slightly acid.

Oa—7 to 19 inches; black sapric material, no fibers present after rubbing, subangular blocky structure, slightly acid.

20e—19 to 45 inches; black sapric material, about 20 percent fibers after rubbing, massive, neutral.

30a—45 to 60 inches; black sapric material, about 10 percent fibers after rubbing, massive, groundwater seeping from profile, neutral.

Colors are for moist soil.

Soil 18 (Pywell Series) with pasture. Very deep gravelly glacial outwash soils are in the far left with a conifer forest.

Soil 18 (Pywell Series)

Family Classification: Euic Typic Borosaprists
Soil Depth: 60 inches or more
Drainage Class: very poorly drained
Parent Material: organic material
Average Annual Precipitation: about 25 inches
Average Annual Air Temperature: about 45 degrees F.
Average Frost-Free Season: about 95 days
Elevation: 1,700 to 2,700 feet
Topography: nearly level bottomlands
Habitat Type: unclassified
Common Native Vegetation: willow, thinleaf alder, sedge, rush, and pyramid spirea
Occurrence in Idaho: northern part, mainly in Benewah, Bonner, Boundary, and Kootenai counties
Land Use: nonirrigated cropland

The Prominent Characteristics of This Soil Are: organic soil; very poorly drained

The Pywell soils are peaty soils that developed in a very wet environment. These soils are totally unlike any mineral soil in Idaho, in that there is little or no mineral soil material throughout the profile. In places there are thin layers of volcanic ash which came from volcanic eruptions in western Washington and western Oregon. Other than that, they formed wholly in the remains of water-tolerant plants. Partially decomposed plant stems and leaves can still be seen, especially at lower depths.

The normal soil development processes common to mineral soils, such as weathering of the parent material to clay and the moving of clay, soluble salts, and carbonates to lower depths, does not exist. That is why the genetic labels given to soil horizons like A and Bt are not used.

These soils are saturated, or nearly saturated, with water most of the year unless they have been drained. The water table is at or near the surface most of the year.

When these soils are diked and drained in order to allow planting and harvesting of crops, faster decomposition of the surface soil takes place. This destroys the soil. Subsidence, or the lowering of the ground level, occurs, depending upon the design of the drainage system. Proper management, by maintaining a high water table as long as possible, is necessary to prevent destruction of these soils.

Also, if the water table is lowered to the point that the surface layer dries out it can catch on fire. Once started it may burn for many months or years despite all efforts to put it out. The area thus burned can revert to a lake.

When pastures are developed on these soils, high forage yields are obtained. Crops commonly grown are spring wheat, oats, and barley, all of which produce good yields. Undeveloped areas of these soils provide excellent habitat for wetland wildlife.

All other uses of these soils are greatly restricted. These soils will not support buildings and roads without expensive preparation. The soils have a high water table and are subject to flooding, Septic systems are not practical.

Soil 19 (Vay Series)

Bw—0 to 6 inches*; dark yellowish brown gravelly silt loam, granular structure, slightly acid.

Bs—6 to 19 inches; yellowish brown gravelly silt loam, subangular blocky and granular structure, slightly acid.

2C1—19 to 44 inches; very pale brown very gravelly coarse sandy loam, subangular blocky structure, few cobblestones, medium acid.

2C2—44 to 61 inches; very pale brown very gravelly coarse sandy loam, massive, medium acid.

*Thin A horizon at surface.

Soil 19 (Vay Series) on steep mountain slopes.

Soil 19 (Vay Series)

Family Classification: medial over loamy-skeletal, mixed Entic Cryandepts

Soil Depth: 40 to 60 inches or more to weathered granite or gneiss

Drainage Class: well drained

Parent Material: volcanic ash mixed with material weathered from granite or gneiss

Average Annual Precipitation: about 40 inches

Average Annual Air Temperature: about 40 degrees F.

Average Frost-Free Season: about 75 days

Elevation: 3,200 to 5,000 feet

Topography: steep and very steep mountains

Habitat Type: *Thuja plicata/Pachystima myrsinites*

Common Native Vegetation: western redcedar, grand fir, Douglas-fir, western larch, western white pine, low Oregon-grape, myrtle pachystima, and American trailplant

Occurrence in Idaho: northern part, mainly in Bonner, Boundary, and Shoshone counties

Land Use: woodland and watershed

The Prominent Characteristics of This Soil Are: highly erodible volcanic ash surface soil; more than 35 percent coarse fragments

The following 14 soils have one thing in common — they have had only minimal soil development in a moist climate. A horizons generally are thin except for **Soil 23 (Roseberry Series).** The Roseberry soil has a thick A horizon with low natural fertility. B horizons have little clay that has been translocated from surface layers or they have had only some chemical reaction of iron producing brown colors.

The Vay soils are valuable woodland soils occurring in mountainous areas of the panhandle of Idaho. They are capable of producing some of the highest annual tree growth in the state. This is due to the cool moist climate and the volcanic ash on the surface which are favorable for conifers.

These soils have weak development. For many thousands of years they formed little more than a thin A horizon containing a high percentage of coarse fragments overlying granitic residuum. The soils were droughty and the vegetation probably grew slowly.

Then came a series of volcanic ashfalls from volcanoes in western Oregon and western Washington. By far the greatest amount came from the violent eruption of Mt. Mazama in southwestern Oregon, about 6,700 years ago. Crater Lake is now where this once mighty mountain stood. The ash fell on all soils in northern Idaho, but was quickly eroded from those soils which did not have a dense cover of trees to hold it in place.

This ashfall greatly improved the water holding capacity of the soils and therefore its potential for increased vegetative growth. Volcanic ash, in contrast to most soil materials, is not only light weight but highly erodible. If this layer is allowed to erode away by improper care of the land, these soils will no longer produce high yields of timber and the regeneration of desirable species will be more difficult.

These soils receive high annual snowfall and are important for dependable watershed yields for power needs and irrigation. They also provide excellent habitat for big game animals such as deer, elk, and bear.

Soil 20 (Bluehill Series)

A—0 to 3 inches; light brownish gray fine sandy loam, granular structure, mildly alkaline.

Bw—3 to 8 inches; pale brown very fine sandy loam, subangular blocky structure, moderately alkaline.

Bk1—8 to 13 inches; pale brown very fine sandy loam, massive, about 55 percent lime coated tuff pieces, strongly alkaline.

Bk2—13 to 25 inches; very pale brown loam, massive, about 90 percent lime coated tuff pieces, strongly effervescent, strongly alkaline.

Soil 20 (Bluehill Series) on rolling hills. The soil in the foreground is in volcanic ash alluvium.

Soil 20 (Bluehill Series)

Family Classification: ashy, mesic Typic Vitrandepts

Soil Depth: 20 to 40 inches to weakly consolidated volcanic ash

Drainage Class: somewhat excessively drained

Parent Material: volcanic ash

Average Annual Precipitation: about 14 inches

Average Annual Air Temperature: about 46 degrees F.

Average Frost-Free Season: about 105 days

Elevation: 4,600 to 6,000 feet

Topography: moderately steep to very steep hills

Habitat Type: *Artemisia tridentata wyomingensis/Oryzopsis hymenoides*

Common Native Vegetation: Wyoming big sagebrush, tall gray rabbitbrush, needleandthread, Indian ricegrass, bluebunch wheatgrass, Sandberg bluegrass, and arrowleaf balsamroot

Occurrence in Idaho: southern part, mainly in Cassia county

Land Use: rangeland

The Prominent Characteristics of This Soil Are: apparent coarse fragments which easily break down; weakly developed subsoil; highly erodible volcanic ash surface soil

This soil is one of the few soils in the state which developed wholly in volcanic ash. Unlike the ash deposits in northern Idaho, which date mostly about 6,700 years ago, or the even more recent ash in and around the Craters of the Moon National Monument, this ash is very old. It is at least several million years old. The ash came from an unknown source and settled into a large body of water in the southern-most part of the state.

Since the ashfall, there has been a very slow uplift of the area. This uplift has resulted in the development of deep drainage systems which has exposed the ash to depths of a thousand feet or more.

With this great thickness the ash compressed into a weakly consolidated mass. It is this hardened character which gives the impression that it is indeed rock. However, it quickly breaks down when crushed. For this reason, this soil cannot be used as a source of gravel for construction purposes or road surfaces.

Like many soils which have steep slopes in a rather dry environment, the development of soil horizons is slow. At the same time, natural geologic erosion has been comparatively rapid keeping the soil forever young.

This soil developed somewhat like **SOIL 10 (Owyhee Series),** except in a very different kind of parent material. There was enough precipitation to leach lime from the surface layer, but not enough to wash it completely from the soil profile. Also the subsoil, or B horizons, is weakly developed. Strong structure and clay accumulation is lacking.

The ash has a low bulk density of about 1.0, which is the same as water. It is also very easily eroded.

This soil has been used only as rangeland. It has little potential for any other use because of the steep slopes.

Soil 21

Bw—0 to 6 inches; dark brown silt loam, granular structure, neutral.

Bs—6 to 20 inches; brown silt loam, granular structure, neutral.

2Bwb1—20 to 33 inches; dark grayish brown gavelly silt loam, subangular blocky structure, slightly acid.

2Bwb2—33 to 53 inches; yellowish brown gravelly silt loam, subangular blocky structure, few cobblestones, medium acid.

2Bwb3—53 to 61 inches; yellowish brown gravelly loam, subangular blocky structure, medium acid.

Colors are for moist soil.

Soil 21 is on wooded hills. Soils in the foreground formed in alluvium and have a high water table.

Soil 21

Family Classification: medial over loamy, mixed, frigid Typic Vitrandepts
Soil Depth: 60 inches or more
Drainage Class: well drained
Parent Material: volcanic ash over schist residuum
Average Annual Precipitation: about 45 inches
Average Annual Air Temperature: about 44 degrees F.
Average Frost-Free Season: about 75 days
Elevation: 3,000 to 4,000 feet
Topography: steep and very steep foothills
Habitat Type: *Thuja plicata/Pachystima myrsinites*
Common Native Vegetation: western redcedar, grand fir, Douglas-fir, western white pine, and myrtle pachystima
Occurrence in Idaho: northern part, mainly in Clearwater and Latah counties
Land Use: woodland and watershed

The Prominent Characteristics of This Soil Are: highly erodible volcanic ash surface soil; high available water holding capacity

This soil is somewhat like **Soil 19 (Vay Series).** It occurs in mountainous terrain and developed in two kinds of parent material — volcanic ash over old weathered schist. Vay soils differ in having granite or gneiss residuum underlying similar volcanic ash. Notice the apparent difference in color between these two soils. This is due mostly to the moist versus dry condition of the soils at the time they were photographed.

Both soils received about the same amount of ash, mostly from the violent and massive eruption of Mt. Mazama in southwestern Oregon, about 6,700 years ago. Weathering since this major ashfall has released iron which has oxidized giving the brown color to the surface soil. This is typical of many soils in the northern part of the state. For example, see, in addition to the Vay soils, **Soil 24** and **Soil 29 (Bonner Series).** The brown color is especially noticeable when the soils are moist as in the photo of **Soil 21.**

The volcanic ash is mostly silt size and is light weight and fluffy when dry. Its main value is that it holds lots of water for plant growth. Plants commonly concentrate their roots in the upper foot or so even in very deep, well-drained soils, but this soil has more roots than usual in the ash. There has been little or no mixing of lower material into the ash in this soil as compared to other soils with Mazama ash.

Tree growth is excellent on this soil, not only due to the high moisture holding capacity, but the cool moist climate which conifers prefer. Production would doubtless be reduced if the fragile ash were allowed to erode.

The underlying weathered schist is also erodible so road building and maintenance must be carefully completed.

Grazing by livestock is only a temporary use after logging or forest fire. In a few years the tree canopy becomes complete and foragable species are greatly reduced.

Soil 22 (Moonville Series)

A—0 to 6 inches; dark brown gravelly sandy loam, granular structure, neutral.

Bw1—6 to 32 inches; dark yellowish brown very gravelly sandy loam, subangular blocky structure, neutral.

Bw2—32 to 51 inches; dark yellowish brown and black very gravelly sand, single grain, many krotovinas, neutral.

C—51 to 60 inches; black extremely gravelly sand, single grain, slightly acid.

Soil 22 (Moonville Series) is in the foreground under sagebrush. Background is a recent lava flow a few hundred years, or more, old. A cinder cone is in the distance.

Soil 22 (Moonville Series)

Family Classification: cindery, frigid Mollic Vitrandepts

Soil Depth: 60 inches or more

Drainage Class: somewhat excessively drained

Parent Material: volcanic ash and cinders

Average Annual Precipitation: about 14 inches

Average Annual Air Temperature: about 44 degrees F.

Average Frost-Free Season: about 75 days

Elevation: 4,800 to 6,000 feet

Topography: nearly level to moderately sloping plains

Habitat Type: *Artemisia tridentata vaseyana/Festuca idahoensis*

Common Native Vegetation: mountain big sagebrush, antelope bitterbrush, Idaho fescue, and bluebunch wheatgrass

Occurrence in Idaho: south-central part, mainly in Butte county

Land Use: rangeland

The Prominent Characteristics of This Soil Are: highly erodible volcanic ash surface soil; cinders in substratum; rapid permeability

This soil developed in a recent deposition (about 2,000 years ago) of volcanic ash and cinders. Unlike the silt-sized volcanic ash of northern Idaho, which was carried airborne more than 300 miles, this material was ejected from nearby volcanic vents mostly within or near the Craters of the Moon National Monument, and is mostly sand and gravel size.

The lower part of the soil below 38 inches is relatively pure volcanic cinders which are excavated and used locally to surface farm roads. They look and feel like clinkers from a coal burning furnace. The textural descriptions given on the facing page, such as very gravelly sand, refers to the sizes of the cinders and ash.

As the volcanic activity subsided, more and more ash was ejected on top of the cindery base. The ash has undergone only minimal development due to the short time since deposition as well as being in a rather dry climate where changes take place more slowly than in a more moist one. The only development has been a modest accumulation of organic matter in the surface soil and chemical changes in the subsoil giving it the brown colors.

The volcanic ash in the surface soil has low bulk density (0.75 to 0.85) and is very erodible if the vegetation is removed either by overgrazing or by an attempt to cultivate it.

Rapid permeability in the lower part of this soil could allow contamination of the ground water if septic tank filter fields were used in this soil.

Soil 23 (Roseberry Series)

Ag1—0 to 6 inches; very dark gray coarse sandy loam, mottles, granular structure, medium acid.

Ag2—6 to 13 inches; very dark gray coarse sandy loam, mottles, subangular blocky structure, strongly acid.

Ag3—13 to 19 inches; dark brown loamy coarse sand, mottles, subangular blocky and granular structure, medium acid.

Cg—19 to 31 inches; brown loamy coarse sand, mottles, single grain, medium acid.

Colors are for moist soil.

Water table.

Soil 23 (Roseberry Series) is in pasture in the foreground. Steep slopes in the background have well drained soils formed in granite.

Soil 23 (Roseberry Series)

Family Classification: sandy, mixed Humic Cryaquepts

Soil Depth: 60 inches or more

Drainage Class: poorly drained

Parent Material: glacial outwash

Average Annual Precipitation: about 23 inches

Average Annual Air Temperature: about 40 degrees F.

Average Frost-Free Season: about 70 days

Elevation: 3,800 to 5,000 feet

Topography: nearly level plains

Habitat Type: unclassified

Common Native Vegetation: willow, sedge, rush, tufted hairgrass, cinquefoil, common camas, and valerian

Occurrence in Idaho: central part, mainly in Valley county

Land Use: hay and pasture

The Prominent Characteristics of This Soil Are: fluctuating water table; thick, very dark colored surface layer; sandy textures; acid reaction

Roseberry soils developed in sandy glacial outwash deposits. There were small glaciers in the nearby granitic mountains during the ice age. At times, major ice melting and much local flooding took place, filling streambeds and larger valleys with mostly coarse sandy material. In the areas where these soils occur, this very thick sandy deposit spread over older, less permeable, silty lake-laid sediments which now hold up a high water table for much of the year.

The wet environment produced a rank growth of water tolerant grasses, forbs, and shrubs. This quickly formed organic matter which darkened the surface soils. Very little weathering has taken place on these granitic sands. The little amount of calcium, magnesium, sodium, and potassium that did weather quickly leached out of the soil leaving mostly hydrogen to be held onto the clay and organic matter particles. The soil then remains quite acid with low natural fertility.

Most areas of these soils are used for pasture. Many cattle are trucked from southern Idaho to graze these pastures all summer. The climate is cool with a short frost-free season. Oats are sometimes grown, mainly as a means of preparing the land for reseeding a pasture which has been invaded with less desirable forage plants.

These soils are poor locations for construction of roads and buildings. The high water table is a serious problem that would require a lot of expense to overcome. Basements would be generally impractical to build. Buried steel and concrete pipe, if left untreated, is susceptible to corrosion due to the high acidity of these poorly drained soils.

Soil 24

Bw—0 to 6 inches; reddish yellow cobbly silt loam, granular structure, slightly acid.

Bs—6 to 18 inches; reddish yellow very cobbly very fine sandy loam, granular structure, slightly acid.

2BC—18 to 33 inches; light yellowish brown very gravelly sandy loam, subangular blocky structure, slightly acid.

2C—33 to 56 inches; pale brown extremely gravelly coarse loamy sand, single grain, medium acid.

Soil 24 in a small clear-cut area showing many glacier-deposited granite boulders on the surface. Far background shows glacial scouring.

Soil 24

Family Classification: loamy-skeletal, mixed Andic Cryochrepts

Soil Depth: 60 inches or more

Drainage Class: well drained

Parent Material: volcanic ash mixed with glacial till

Average Annual Precipitation: about 50 inches

Average Annual Air Temperature: about 38 degrees F.

Average Frost-Free Season: 50 days or less

Elevation: 4,800 to 6,600 feet

Topography: steep and very steep mountains

Habitat Type: *Abies lasiocarpa/Xerophyllum tenax*

Common Native Vegetation: subalpine fir, Engelmann spruce, western larch, big blueberry, and common beargrass

Occurrence in Idaho: northern part, mainly in Bonner and Boundary counties

Land Use: woodland and watershed

The Prominent Characteristics of This Soil Are: highly erodible volcanic ash surface soil; more than 35 percent rock fragments; boulders on the surface

This soil occurs in the granitic mountains of the northernmost part of Idaho. The continental glacier which entered northern Idaho from Canada was up to a mile thick. Only the highest peaks escaped the scouring which took place. Some of the glacial debris, called glacial till, left by the ice remained on the mountain slopes. Granitic boulders are common.

This soil began its development in this glacial till. The cold temperatures, however, allowed little weathering to take place. The effect of latitude and nearness to arctic air flow on temperatures in this part of the state is in contrast to temperatures at comparable elevations in southern Idaho. Temperatures in northern Idaho are much cooler. For example, see **Soil 48 (Hymas Series).**

Soil 24 later received about 7 to 14 inches of ash from several volcanoes in western Oregon and western Washington. But, like **Soil 19 (Vay Series)** and **Soil 29 (Bonner Series),** the greatest amount came about 6,700 years ago from the eruption of Mt. Mazama in southwestern Oregon. Crater Lake is in the collapsed remains of this mountain. The total amount of ash thrown out of this mountain at that time was about 40 times more than was ejected by Mt. St. Helens in 1980!

Weathering of the ash during the last 6,700 years has produced oxidized iron giving it a distinctive color. It is not only light weight when dry and holds lots of water for plant growth, but it is highly erodible on these mountain slopes. Logging operations have to be carefully done. If the ash is completely removed by erosion, not only will unwanted sedimentation occur, but the productive capacity of the soil will be lowered significantly. The soil material below the volcanic ash, having a higher percentage of rock fragments, has a low capacity for holding water.

Boulders at the surface hinders timber harvest operations.

This is a valuable watershed soil. Very deep snow is common most years. Wildlife habitat is good for summer grazing of big game animals, especially in recently logged areas.

Soil 25

A—0 to 7 inches; dark grayish brown gravelly loam, granular structure, neutral.

Bw—7 to 27 inches; light brownish gray very gravelly loam, subangular blocky structure, slightly acid.

C1—27 to 42 inches; very pale brown extremely cobbly sandy loam, massive, slightly acid.

C2—42 to 61 inches; white extremely gravelly sandy loam, massive, slightly acid.

Soil 25 on steep mountain slopes. Bare areas are scree.

Soil 25

Family Classification: loamy-skeletal, mixed Dystric Cryochrepts

Soil Depth: 60 inches or more

Drainage Class: well drained

Parent Material: quartzite colluvium

Average Annual Precipitation: about 24 inches

Average Annual Air Temperature: about 36 degrees F.

Average Frost-Free Season: 50 days or less

Elevation: 6,000 to 8,000 feet

Topography: steep and very steep mountains

Habitat Type: *Abies lasiocarpa/Xerophyllum tenax*

Common Native Vegetation: Douglas-fir, lodgepole pine, Engelmann spruce, subalpine fir, big blueberry, and common beargrass

Occurrence in Idaho: east-central part, mainly in Lemhi county

Land Use: woodland and watershed

The Prominent Characteristics of This Soil Are: thin, dark colored surface layer; weakly developed subsoil; more than 35 percent coarse fragments; low available water holding capacity

This soil occurs in mountainous regions having one of the coolest climates in Idaho. The presence of subalpine fir and beargrass which survive in such a climate is the clue to the cool temperatures. This cool climate not only adversely affects all plants which have only a brief summer in which to grow, but the weathering of the parent material is slow in contrast to a soil like **Soil 43 (Gwin Series),** which has a much warmer temperature.

Soil 25 developed only a thin A horizon. The Bw horizon is very weakly developed, having only some blocky structure with no detectable increase in clay, which is common in well developed B horizons such as those in the Gwin soils.

Soil 25 is very droughty. It contains a high percentage of rock fragments which is typical of many soils occurring in mountainous topography. As a result the total available water capacity is low. The similar **Soil 19 (Vay Series)** received a layer of volcanic ash which makes its capacity to hold water much greater than **Soil 25.**

Low summer precipitation coupled with the low water holding capacity limits tree growth. This and the rugged terrain of this soil makes it only moderately valuable as a wood producing soil. Most of the value of this soil comes from the watershed yield for irrigation and power development downstream.

Wildlife habitat is excellent for big game animals. Hiking trails in this scenic part of Idaho are developed on this soil.

Soil 26

A—0 to 7 inches; brown loam, granular structure, few gravel, medium acid.

Bw1—7 to 15 inches; reddish brown loam, granular structure, few gravel, medium acid.

Bw2—15 to 44 inches; reddish brown loam, subangular blocky structure, about 15 percent cobblestones, slightly acid.

Bw3—44 to 56 inches; reddish brown cobbly loam, massive, slightly acid.

Soil 26 across the lake in woodland.

Soil 26

Family Classification: fine-loamy, mixed, Dystric Cryochrepts

Soil Depth: 60 inches or more

Drainage Class: well drained

Parent Material: basalt residuum

Average Annual Precipitation: about 30 inches

Average Annual Air Temperature: about 39 degrees F.

Average Frost-Free Season: about 60 days

Elevation: about 5,000 feet

Topography: gently sloping to steep uplands

Habitat Type: *Abies grandis/Spiraea betulifolia*

Common Native Vegetation: grand fir, Douglas-fir, lodgepole pine, snowbrush ceanothus, willow, pine reedgrass, and pyramid spirea

Occurrence in Idaho: central part, mainly in Adams county

Land Use: woodland and watershed

The Prominent Characteristics of This Soil Are: weakly developed subsoil; reddish-colored subsoil

This colorful soil resembles some of the reddish-colored soils of the tropics such as in parts of Hawaii. Like the soils in Hawaii, this soil developed wholly in basalt residuum. It is one of the few remaining soils in Idaho which existed during the Pliocene epoch over 3 million years ago. All other soils of that era have been lost by erosion with new parent materials being formed and influenced by completely different climates.

During the Pliocene the climate here was tropical or subtropical with continuously warm moist conditions. This caused a different kind of weathering to take place than is common in Idaho today.

This soil long ago was not only deeply weathered but nearly completely weathered, leaving mainly a mixture of quartz, kaolinite, and free oxides, without clearly marked soil horizons. Iron oxide gives the soil its reddish color.

The ancient soil was completely weathered. Even though the climate has changed to a more temperate one, which is favorable for the formation and movement of a mixture of kinds of clays, this soil has not been able to develop further. This is because of the presence of large amounts of iron, aluminum, and other metallic oxides and kaolinite. Oxides and kaolinite are the final products in the weathering process.

The agricultural potential of this soil, even on the gently sloping areas, is limited to forest products because the growing season is too short and cool for cultivated crops.

This soil is well suited as a watershed for much needed water downstream and as a wildlife habitat. The more gentle slopes are being used more and more as winter recreational areas for cross country skiing and snowmobiling.

Soil 27

Bw—0 to 11 inches; yellowish brown sandy loam, granular structure, slightly acid.

Bs—11 to 22 inches; yellowish brown sandy loam, subangular blocky structure, neutral.

C—22 to 40 inches; light yellowish brown gravelly loamy sand, single grain, neutral.

Cr—40 to 63 inches; light brown and brown weathered gneiss.

Soil 27 with western redcedar, Douglas-fir, and grand fir in steep mountainous slopes.

Soil 27

Family Classification: coarse-loamy, mixed, frigid Andic Dystrochrepts

Soil Depth: 40 to 60 inches to weathered granite or gneiss

Drainage Class: well drained

Parent Material: volcanic ash mixed with material weathered from granite or gneiss

Average Annual Precipitation: about 40 inches

Average Annual Air Temperature: about 42 degrees F.

Average Frost-Free Season: about 75 days

Elevation: 3,000 to 4,400 feet

Topography: steep and very steep mountains

Habitat Type: *Thuja plicata/Pachystima myrsinites*

Common Native Vegetation: western redcedar, grand fir, Douglas-fir, myrtle pachystima, snowberry, rose, and queencup beadlily

Occurrence in Idaho: northern part, mainly in Clearwater county

Land Use: woodland and watershed

The Prominent Characteristics of This Soil Are: highly erodible volcanic ash surface soil; easily rippable bedrock; steep slopes

This soil is similar to **Soil 19 (Vay Series)** in having volcanic ash mixed with material weathered from granite or gneiss, in having mountainous topography, and having some of the highest annual tree growth in the state.

The ash cap on this soil came from the same origin as many other soils in northern Idaho, such as **Soil 24** and **Soil 29 (Bonner Series).** Volcanic ash came from several volcanic vents in western Oregon and western Washington, but the greatest of all came from Mt. Mazama in southwestern Oregon, about 6,700 years ago. Crater Lake is the remains of that violent explosion. A little more than half as much ash fell on this soil as on the Vay soils. The ash is important since it improves the moisture-holding capacity. This fragile material could easily be lost by erosion if good forest management practices are not carried out. The result would be permanent lowering of the productive capacity of this soil as well as increasing the sedimentation downstream.

The underlying material of this soil has only a moderate amount of gravel in contrast to the very gravelly Vay soils. It does, however, have a high percentage of sand so the moisture-holding capacity of this part of the soil is low.

The lower part of the soil profile is weathered bedrock. This will eventually become part of the soil, but for now it allows no root development except in a few cracks. It is however, easily rippable by machinery in excavating or road building projects.

The steep slopes largely restrict this soil to woodland use. Wildlife habitat is good for big game animals, especially if there are cleared areas nearby which have allowed the growth of shrubs.

This soil has favorable infiltration rates and rapid permeability allowing for groundwater recharge. Runoff is nearly non-existent from this soil under natural vegetative conditions. Stream flow, even from small streams, continues all summer even though little precipitation occurs during this period.

Soil 28 (Moonville Variant)

A—0 to 7 inches; dark yellowish brown loam, granular structure, few cinders, neutral.

Bw—7 to 30 inches; yellowish brown gravelly loam, subangular blocky structure, mildly alkaline.

Bk—30 to 51 inches; very pale brown loam, massive, common 2 to 5 inch diameter krotovinas, strongly effervescent, moderately alkaline.

Soil 28 (Moonville Variant) with sagebrush.

Soil 28 (Moonville Variant)

Family Classification: coarse-loamy, mixed, frigid Andic Xerochrepts

Soil Depth: 60 inches or more

Drainage Class: well drained

Parent Material: volcanic ash and cinders

Average Annual Precipitation: about 14 inches

Average Annual Air Temperature: about 42 degrees F.

Average Frost-Free Season: about 80 days

Elevation: 5,000 to 6,000 feet

Topography: nearly level and gently sloping plains

Habitat Type: Artemisia tridentata vaseyana/Festuca idahoensis

Common Native Vegetation: mountain big sagebrush, antelope bitterbrush, Idaho fescue, bluebunch wheatgrass, Thurber needlegrass, Sandberg bluegrass, lupine, and arrowleaf balsamroot

Occurrence in Idaho: southern part, mainly in Butte county

Land Use: rangeland

The Prominent Characteristics of This Soil Are: highly erodible volcanic ash surface soil; high available water holding capacity

This soil, like **Soil 22 (Moonville Series)**, developed in volcanic ash and cinders which came from volcanic vents in and near the Craters of the Moon National Monument during the last few thousand years. However, since **Soil 28** is farther from the source the ash is smaller sized and contains less cinders. The gravelly textures described in the soil profile description refer to the presence of cinders, not hard rock material.

The volcanic ash of **Soil 28 (Moonville Variant)**, along with the cinders, contrasts with the ash of northern Idaho. The northern Idaho ash is not only older but it traveled hundreds of miles by air so it is all silt size with no cinders.

This very deep soil, having a high percentage of volcanic ash, has a high available water capacity. Although this soil is now used only as rangeland, it could be cultivated should water become available. However, careful irrigation water management would be necessary to control erosion on this fragile material.

This soil has undergone only minimal development since the parent material has been in place, due not only to the limited time available, but the rather dry environment which keeps soil development at a slow pace.

Lime has been removed from the upper parts of the soil and is now in the Bk horizon. The krotovinas appearing in the photograph as darker colored spots in this zone are the result of small animal diggings which mix in darker soil from the upper layers. This is like the krotovinas seen in **Soil 37.**

The surface layer has accumulated very little organic matter and the subsoil has only blocky structure with no weathering of the parent material to clay.

Soil 29 (Bonner Series)

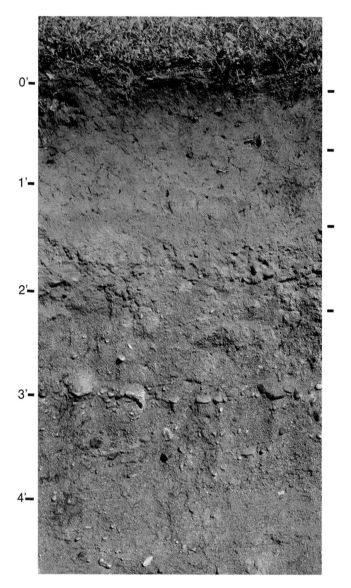

Bw—0 to 7 inches; dark yellowish brown silt loam, granular structure, few gravel, slightly acid.

Bs—7 to 15 inches; dark yellowish brown gravelly silt loam, blocky structure, medium acid.

2BC—15 to 25 inches; very pale brown gravelly coarse sandy loam, blocky structure, strongly acid.

2C—25 to 58 inches; multi-colored very gravelly coarse sand, single grain, medium acid.

Colors are for moist soil.

Soil 29 (Bonner Series) is in pastures. The low mountains have been scoured by glaciers.

Soil 29 (Bonner Series)

Family Classification: coarse-loamy over sandy or sandy-skeletal, mixed frigid Andic Xerochrepts
Soil Depth: 60 inches or more
Drainage Class: well drained
Parent Material: volcanic ash over glacial outwash
Average Annual Precipitation: about 30 inches
Average Annual Air Temperature: about 43 degrees F.
Average Frost-Free Season: about 110 days
Elevation: 2,000 to 3,000 feet
Topography: nearly level and gently sloping terraces
Habitat Type: *Abies grandis/Pachystima myrsinites*
Common Native Vegetation: grand fir, Douglas-fir, ponderosa pine, western larch, lodgepole pine, myrtle pachystima, rose, pine reedgrass, sedge, and longtube twinflower
Occurrence in Idaho: northern part, mainly in Bonner, Boundary, and Kootenai counties
Land Use: woodland and nonirrigated and irrigated cropland

The Prominent Characteristics of This Soil Are: low available water holding capacity; gravel and sand substratum

The Bonner soils occur in areas which were repeatedly washed and filled by flood waters from the Lake Missoula floods. The last time this occurred was about 13,000 years ago. The glacial outwash materials are from a variety of rocks including granite, shale, siltstone, sandstone, quartzite, limestone, and argillite. These materials were scoured and carried by fingers of huge glaciers moving from Canada into northern Idaho.

The deposits of mostly sand and gravel are somewhat layered or stratified because of varying speeds of the flood waters.

For several thousand years, or until this outwash was blanketed by volcanic ash, this soil must have supported only sparse vegetation because the gravelly lower part has a very low available water capacity.

There were minor ash deposits during this interval of time, but by far the largest ash fallout occurred about 6,700 years ago. This was when, with a huge explosion, Mt. Mazama covered northern Idaho with a foot or more of ash. Crater Lake in southwestern Oregon is a result of this explosion.

Since that time there has been some mixing of this ash with the underlying sand and gravel to form the Bonner soils. This mixing resulted from freezing and thawing action, uprooting of trees, and small animal activity.

Now 14 to 24 inches of the upper part of the Bonner soils contains up to 60 percent volcanic ash. The bulk density of this part of the soil is about 0.85 to 0.95, which is lighter than an equal volume of water. The ash has a high available moisture holding capacity which is a big improvement over the old underlying soil.

This soil has a wide variety of uses. Timber yields are good, but tree seedlings receive severe competition from brush after harvest.

Bonner soils are too droughty during summer months for good production of dryland crops. Irrigation improves yields substantially. Crops include spring wheat, oats, barley, hay, and pasture. Christmas trees and nursery stock are also being grown.

There are many housing subdivisions on these soils. The rapid or very rapid permeability of the lower parts of this soil may cause groundwater pollution under heavy use of septic systems.

Excellent sources of sand and gravel can be found under these soils.

Soil 30 (Oxford Series)

Ap—0 to 8 inches; light brown silty clay, granular structure, strongly effervescent, mildly alkaline.

Bw1—8 to 15 inches; light reddish brown silty clay, subangular blocky structure, strongly effervescent, mildly alkaline.

Bw2—15 to 47 inches; light reddish brown silty clay, prismatic structure, extremely hard when dry, few slickensides, strongly effervescent, mildly alkaline.

By—47 to 62 inches; light reddish brown silty clay, subangular blocky structure, many calcium sulfate crystals, strongly effervescent, mildly alkaline.

Soil 30 (Oxford Series) with fall-plowed winter wheat stubble.

Soil 30 (Oxford Series)

Family Classification: fine, montmorillonitic, frigid Vertic Xerochrepts

Soil Depth: 60 inches or more

Drainage Class: well drained

Parent Material: lacustrine

Average Annual Precipitation: about 17 inches

Average Annual Air Temperature: about 44 degrees F.

Average Frost-Free Season: about 95 days

Elevation: 4,700 to 5,100 feet

Topography: gently sloping to moderately steep dissected lakebed terraces

Habitat Type: *Artemisia tridentata vaseyana/Agropyron spicatum*

Common Native Vegetation: mountain big sagebrush, antelope bitterbrush, bluebunch wheatgrass, and arrowleaf balsamroot

Occurrence in Idaho: southeastern part, mainly in Franklin and Bannock counties

Land Use: nonirrigated cropland

The Prominent Characteristic of This Soil Is: clayey subsoil with high shrink-swell

The Oxford soils developed in lakebed sediments of old Lake Bonneville. Since the lake drained during the Bonneville flood about 14,000 years ago, the area has been dissected by geologic erosion into a somewhat hilly topography. The present Great Salt Lake, 50 miles to the south, is 4,200 feet above sea level and is the remnant of the much larger Lake Bonneville.

There has been little soil development. The reddish color comes from the color of the surrounding shale, which eroded and formed sediment in this part of the lake, rather than from weathering. The sediments were clayey, having been settled out of deep still water. The soil has lime throughout. Without the removal of lime, at least from the upper part, there is little chance of movement of more clay into the subsoil.

The kind of clay originally deposited was mostly the kind which swells when moist and shrinks when dry, similar to **Soil 54 (Boulder Lake Series).** This has caused mixing to occur in the upper 2 or 3 feet. This alternate shrinking and swelling easily destroys structures such as foundation walls, sidewalks, or streets.

All of this soil is being cropped, producing mostly wheat and barley without irrigation. The clayey surface soil can be a problem in cultivation. If cultivated when the soil is too wet or too dry, it will cause poor tilth.

Septic systems, even on gentle slopes, would not function properly because of the very slow permeability of the subsoil.

Soil 31 (McCall Series)

A—0 to 8 inches; dark grayish brown very stony sandy loam, granular structure, medium acid.

Bw1—8 to 20 inches; brown very stony sandy loam, subangular blocky structure, slightly acid.

Bw2—20 to 36 inches; pale brown very stony sandy loam, subangular blocky structure, slightly acid.

C—36 to 64 inches; light brownish gray very stony coarse sand, single grain, slightly acid.

Soil 31 (McCall Series) cleared of trees and in pasture. Boulders deposited by glaciers are common on these soils.

Soil 31 (McCall Series)

Family Classification: loamy-skeletal, mixed Typic Cryumbrepts

Soil Depth: 60 inches or more

Drainage Class: somewhat excessively drained

Parent Material: glacial till

Average Annual Precipitation: about 25 inches

Average Annual Air Temperature: about 39 degrees F.

Average Frost-Free Season: about 70 days

Elevation: 4,900 to 5,300 feet

Topography: gently sloping to steep uplands

Habitat Type: *Abies grandis/Vaccinium membranaceum*

Common Native Vegetation: grand fir, Douglas-fir, ponderosa pine, big blueberry, mallow ninebark, snowberry, pine reedgrass, sedge, and heartleaf arnica

Occurrence in Idaho: central part, mainly in Valley county

Land Use: woodland, hay, and pasture

The Prominent Characteristics of This Soil Are: more than 35 percent rock fragments; boulders on the surface; very low available water-holding capacity

McCall soils occur on glacial moraines. Large boulders moved by ice, as seen in the landscape photograph are common. Glaciers move rock material of all sizes from boulders down to silt-sized rock flour. Glaciers in this part of Idaho were local in nature and moved relatively short distances. That is why some of the rock fragments in these soils are only partially rounded by the grinding action of the glaciers. This contrasts with the huge lobes of the continental glacier which moved long distances from Canada into northern Idaho.

The formation of Payette Lake, in Valley county of central Idaho, was the result of a local glacier deepening an existing valley and dropping its load of rock debris forming a dam.

Since these soils developed in a thick deposit of glacial till, the depth to bedrock is far below the usual soil depth of 5 feet. The total available water-holding capacity is limited by the high percentage of rock fragments as well as the moderately coarse and coarse textured subsoil and substratum. The total of less than 3 inches available water is in contrast to a soil like **Soil 50 (Palouse Series),** which has about 12 inches of available water-holding capacity. The very low available water capacity of McCall soils limits the growth of trees.

Some areas of these soils have been cleared and used as pasture. The presence of boulders and stones, as well as the cool short growing season, generally prohibits the growing of cultivated crops.

McCall soils are somewhat like **Soil 24** in that they both developed in glacial till. The important difference is that **Soil 24** received a cover of volcanic ash on the surface. This greatly improved the moisture-holding capacity resulting in much greater productive potential for trees.

Soil 32

A1—0 to 12 inches; very dark brown loam, granular structure, few gravel, strongly acid.

A2—12 to 18 inches; dark brown loam, granular structure, few gravel, medium acid.

Bw1—18 to 28 inches; dark yellowish brown loam, subangular blocky structure, few gravel, medium acid.

Bw2—28 to 39 inches; yellowish brown very stony loam, subangular blocky structure, medium acid.

C—39 to 50 inches; pale brown very gravelly loamy sand, single grain, strongly acid.

Soil 32 is in the foreground and in the grassy area in the distance. Soils under the trees have thin A horizons.

Soil 32

Family Classification: coarse-loamy, mixed Andic Cryumbrepts

Soil Depth: 60 inches or more

Drainage Class: well drained

Parent Material: volcanic ash mixed with material weathered from granite or gneiss

Average Annual Precipitation: about 50 inches

Average Annual Air Temperature: about 35 degrees F.

Average Frost-Free Season: 50 days or less

Elevation: 5,400 to 7,000 feet

Topography: steep and very steep mountains

Habitat Type: *Abies lasiocarpa/Xerophyllum tenax*

Common Native Vegetation: common beargrass, sedge, big blueberry, mock azalea, white spirea, and lupine

Occurrence in Idaho: northern part, mainly from Boundary to Idaho counties

Land Use: watershed

The Prominent Characteristics of This Soil Are: steep slopes; acid reaction; highly erodible volcanic ash surface soil

This soil is on south-facing ridges in the mountains of northern Idaho. Areas of this soil can easily be seen from a distance because they are covered with a dense stand of grass, mainly beargrass, surrounded by a forest, mostly subalpine fir.

Wild fires were common in these wilderness areas before fire fighting techniques were established in the recent past. Trees were able to reestablish themselves in most areas, except some of the south-facing ridges at the higher elevations. This is probably due to the climate as well as the soil. These exposed ridges are cold and windy and the soil is droughty.

There is always fierce competition for survival between grasses and trees in areas that are marginal for trees. Grasses will take over if given a chance. That chance probably occurred a few hundred years ago when a stand of subalpine fir was burned.

Eventually subalpine fir will probably return but in the meantime grasses are dominant. They have been dominant long enough to darken the surface layer deeper than is common under nearby stands of subalpine fir.

There is an abundance of grass roots in the surface layer — more than in any other soil described in this book. Decomposition is particularly slow in this cool environment and has resulted in an acid soil with low natural fertility.

A few inches of volcanic ash from the Mt. Mazama eruption about 6,700 years ago, which is common in soils like **Soil 19 (Vay Series)** and **Soil 21,** has been mixed with granite or gneiss residuum from below. This thin layer of ash is highly erodible. However, these steep slopes are generally well covered with dense stands of grasses which protect this fragile surface.

The northwest is highly dependent on this kind of soil to provide watersheds that collect deep snow for later release for power production and irrigation.

Paths and trails are common on this soil which provide access to the beautiful mountain scenery.

Soil 33 (Southwick Series)

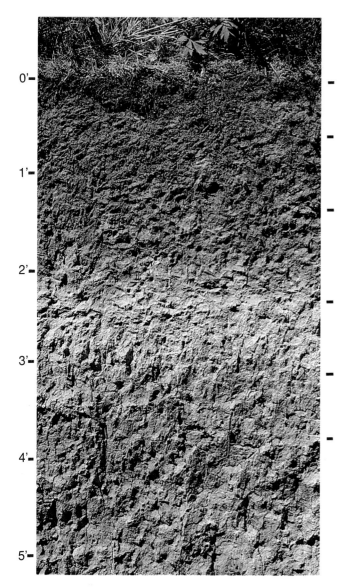

A1—0 to 7 inches; dark brown silt loam, granular structure, slightly acid.

A2—7 to 15 inches; grayish brown silt loam, subangular blocky structure, slightly acid.

Bw—15 to 28 inches; brown silt loam, subangular blocky structure, neutral.

Ec—28 to 37 inches; light gray silt loam, subangular blocky structure, iron concretions, slightly acid.

BE—37 to 45 inches; brown silty clay loam, angular blocky structure, slightly acid.

Btb—45 to 63 inches; yellowish brown silty clay loam, prismatic structure, very hard when dry, neutral.

Soil 33 (Southwick Series) with a crop of dry peas.

Soil 33 (Southwick Series)

Family Classification: fine-silty, mixed, mesic Argiaquic Xeric Argialbolls)
Soil Depth: 60 inches or more
Drainage Class: moderately well drained
Parent Material: loess
Average Annual Precipitation: about 23 inches
Average Annual Air Temperature: about 46 degrees F.
Average Frost-Free Season: about 115 days
Elevation: 2,300 to 3,500 feet
Topography: gently sloping to hilly uplands
Habitat Type: *Pinus ponderosa/Symphoricarpos albus*
Common Native Vegetation: ponderosa pine, snowberry, rose, Saskatoon serviceberry, pine reedgrass, blue wildrye, bluebunch wheatgrass, rough fescue, sedge, and geranium
Occurrence in Idaho: northern part, mainly from Kootenai county to Lewis county
Land Use: woodland and nonirrigated cropland

The Prominent Characteristics of This Soil Are: E soil horizon below the surface soil; slow permeability

This soil and the following 17 soils, although formed in a variety of parent materials and climates, have rather dark-colored surface layers with at least one percent organic matter. They have a relatively high capacity to hold plant nutrients.

Soil 33 (Southwick Series) formed in loess. The loess originated from the same source as did the loess from which **Soil 50 (Palouse Series)** developed. Glacial meltwater, from the continental glaciers which advanced into the United States from Canada, spread over broad areas in south-central Washington. This glacial meltwater carried silty outwash material. Strong winds picked up the silty material and carried it into northern Idaho. Glaciers occurred several times during the Pleistocene age with long periods of time between advancements. During these times soils were developed having A and B horizons. After each advance of ice another loess deposit was added to the landscape and a new soil developed.

The Palouse soils developed wholly in the last major deposit whereas the Southwick soils developed in areas which received only about 2 feet of new loess overlying an older buried soil. The A horizon of this buried soil was altered to an E horizon because a perched water table moving laterally at this point leached out clay. Organic matter, which coated the silt particles of this old A horizon, was also removed, leaving uncoated silt and sand particles, causing this soil horizon to be light colored. Deep new road cuts in the loess of this area show a series of underlying old soils which had formed during times between ice advances.

Most areas of Southwick soils were cleared of ponderosa pine and are now in cultivation. Winter wheat, barley, dry peas, and lentils are grown with good yields. Hay and pasture are also grown. Soil erosion remains the most difficult problem to overcome when these soils are cultivated.

A few areas have remained in woodland which also provides some grazing as well as good habitat for wildlife.

When these soils are used for building development, the perched water table in the Ec horizon and the slow permeability of the Bt horizon can cause problems. Many septic tank filter fields function poorly. Also, unless areas are properly drained with tile, basements can be wet at times.

Soil 34 (Houk Series)

Ap—0 to 10 inches; very dark grayish brown silty clay loam, granular structure, neutral.

Eg—10 to 12 inches; dark gray silt loam, blocky structure, mildly alkaline.

Btg—12 to 19 inches; very dark gray clay, angular blocky structure, mildly alkaline.

Btkg—19 to 30 inches; very dark grayish brown clay loam, few mottles, angular blocky structure, strongly effervescent, moderately alkaline.

Btg—30 to 38 inches; very dark brown clay loam, common mottles, angular blocky structure, moderately alkaline.

2Cg—38 to 47 inches; dark grayish brown sandy loam, common mottles, massive, mildly alkaline.

Colors are for moist soil.

Soil 34 (Houk Series) with spring barley along Camas Creek. Shallow and moderately deep, well drained, medium textured soils derived from rhyolite are in the distance. They are used for range.

Soil 34 (Houk Series)

Family Classification: fine, montmorillonitic, frigid Argiaquic Xeric Argialbolls

Soil Depth: 60 inches or more

Drainage Class: somewhat poorly drained

Parent Material: alluvium

Average Annual Precipitation: about 14 inches

Average Annual Air Temperature: about 41 degrees F.

Average Frost-Free Season: about 85 days

Elevation: 4,800 to 5,400 feet

Topography: nearly level bottomlands and low terraces

Habitat Type: *Artemisia cana/Festuca idahoensis*

Common Native Vegetation: silver sagebrush, willow, Idaho fescue, sedge, hairgrass, basin wildrye, and rush

Occurrence in Idaho: southern part, mainly in Camas county

Land Use: rangeland and nonirrigated and irrigated cropland

The Prominent Characteristics of This Soil Are: grayish colored soil horizons; E soil horizon below the surface soil; fluctuating water table; short frost-free season; flooding hazard

The very dark gray colors, predominant in the Houk soils, are due to their being wet most of the time during their development. This excluded oxygen from entering the soil. Chemical reduction of iron, the opposite of oxidation, took place resulting in the gray colors.

The water table rises to about 30 inches in the spring and lowers to 60 inches in late summer and fall. Note the water seeping out of the soil in the lower part. In some places, subsurface drains have been installed to lower the water table.

The wet nature of these soils provide fair habitat for a wide variety of waterfowl, including ducks, geese, and greater sandhill cranes. Marsh hawk, long-billed curlew, avocet, phalarope, killdeer, and mule deer are also present. Improved habitat for waterfowl would be easy to develop on these soils.

Forage production is high when the natural vegetation is used for rangeland. With proper management, it is easy to maintain the range in good condition.

For most other uses, however, the wet nature and flooding hazard of these soils is a problem. Adapted species that can tolerate the wet conditions are needed when these soils are used for hay or pasture. Good yields are obtained.

Cropland areas are primarily used for spring wheat and barley. Planting is often delayed in spring by the wet soils. In an already short growing season this reduces the suitability for these crops. The tendency to cultivate when the soils are too wet often leads to the development of a plowpan. Chiseling would then be necessary to break up this pan to promote better root penetration and improved aeration.

Soil 35 (Nez Perce Series)

Ap—0 to 7 inches; very dark grayish brown silt loam, granular structure, medium acid.

A—7 to 16 inches; dark grayish brown silt loam, subangular blocky structure, slightly acid.

Ec—16 to 19 inches; pale brown silt loam, subangular blocky structure, iron concretions, neutral.

Btb—19 to 32 inches; dark brown silty clay, prismatic structure, many slickensides, neutral.

Btkb—32 to 55 inches; brown silty clay, prismatic structure, many slickensides, pockets of calcium carbonate, mildly alkaline.

Btb—55 to 62 inches; brown silty clay loam, subangular blocky structure, moderately alkaline.

Soil 35 (Nez Perce Series) with ripening winter wheat. North-facing slopes in the background have soils which lack E horizons.

Soil 35 (Nez Perce Series)

Family Classification: fine, montomorillonitic, mesic Xeric Argialbolls

Soil Depth: 60 inches or more

Drainage Class: moderately well drained

Parent Material: loess

Average Annual Precipitation: about 22 inches

Average Annual Air Temperature: about 46 degrees F.

Average Frost-Free Season: about 120 days

Elevation: 2,800 to 4,000 feet

Topography: gently sloping to hilly uplands

Habitat Type: *Festuca idahoensis/Rosa nutkana*

Common Native Vegetation: Idaho fescue, bluebunch wheatgrass, rose, arrowleaf balsamroot, geranium, and lupine

Occurrence in Idaho: north-central part, mainly in Idaho and Lewis counties

Land Use: nonirrigated cropland

The Prominent Characteristics of This Soil Are: high organic matter in surface soil; E soil horizon below the surface soil; clayey subsoil with high shrink-swell

The Nez Perce soils developed in loess which came from the same source as the loess in which **Soil 50 (Palouse Series)** developed. There are, however, sharp contrasts in these soils.

The plateau where the Nez Perce soils occur is not only a thousand feet higher than where the Palouse soils occur, but is out of the path of the prevailing strong westerly winds. Therefore, there were fewer and much thinner loess deposits where the Nez Perce soils occur. The Palouse soils developed wholly in the most recent major deposit, whereas only the A horizons of the Nez Perce soils developed in this deposit.

Before this recent loess deposit, the Nez Perce soils resembled **Soil 53 (Ager Series)** in that the whole soil was clayey. This clay was the kind with a high shrink-swell. Vertical cracks, like those in the Ager soils, swallowed any thin loess additions to the surfaces, up to 5 to 7 inches. The latest major loess deposit of about 10 to 20 inches arrested this activity. Nearby soils which received less than 5 to 7 inches remain as the Nez Perce soils used to be — clayey to the surface.

The present rate of accelerated soil erosion, caused by cultivation, will inevitably return the Nez Perce soils to their original clayey surface within a few hundred years.

The very dark surface soil is high in organic matter. The cool moist climate was ideal for the rank growth of grasses, producing abundant roots which were responsible for the high organic matter content.

The Ec soil horizon resulted because the freshly-added loess had a more rapid permeability rate than that of the underlying clayey soil. This caused a temporary perched water table which leached clay and organic matter from the Ec horizon. This soil horizon has not only the lightest color but the least amount of clay in the profile.

These soils are so well suited to cultivation that no areas remain in native vegetation. Crops include wheat, barley, dry peas, hay, and pasture. Yields are high, but the hazard of soil erosion is ever present.

When these soils are used for all kinds of building site development, including sanitary facilities, the overriding problem is the clayey subsoil.

Soil 36 (Driggs Variant)

A1—0 to 6 inches; dark brown silt loam, granular structure, few gravel, neutral.

A2—6 to 13 inches; dark brown silt loam, subangular blocky structure, few gravel, neutral.

Bw—13 to 25 inches; yellowish brown very fine sandy loam, prismatic structure, few gravel, neutral.

Bt1—25 to 35 inches; brown gravelly clay loam, subangular blocky structure, neutral.

Bt2—35 to 52 inches; brown very gravelly sandy loam, subangular blocky structure, mildly alkaline.

C—52 to 62 inches; multi-colored extremely gravelly coarse sand, single grain, mildly alkaline.

Soil 36 (Driggs Variant) with winter wheat. Very deep, well drained soils developed in loess are being cultivated in the distance.

Soil 36 (Driggs Variant)

Family Classification: fine-loamy, mixed Argic Cryoborolls

Soil Depth: 60 inches or more

Drainage Class: well drained

Parent Material: alluvium

Average Annual Precipitation: about 15 inches

Average Annual Air Temperature: about 40 degrees F.

Average Frost-Free Season: about 80 days

Elevation: 5,800 to 6,800 feet

Topography: gently and moderately sloping terraces

Habitat Type: *Artemisia tridentata vaseyana/Festuca idahoensis*

Common Native Vegetation: mountain big sagebrush, Idaho fescue, bluebunch wheatgrass, Sandberg bluegrass, arrowleaf balsamroot, and lupine

Occurrence in Idaho: eastern part, mainly in Teton county

Land Use: rangeland and nonirrigated and irrigated cropland

The Prominent Characteristics of This Soil Are: short frost-free season; gravel and sand substratum

Soil 36 (Driggs Variant) occurs where the growing season is not only short but cool. Farmers have learned to cope with these limitations by growing only those crops which thrive under such conditions. Most areas of this soil are now under cultivation and in places are being irrigated. Wheat, barley, and alfalfa are common. Only seed potatoes are grown because the growing season is generally too short to mature commercial potatoes.

The area in which this soil occurs is isolated from the commercial potato growing areas. For this reason it is possible to maintain disease-free plants. Thus most of the seed potatoes used in Idaho are grown on this soil.

Another specialty crop, because of the cool temperature, is peas. They are picked in the pod and shipped in iced railroad cars to eastern markets that demand a high quality green pea.

Areas of this soil, lacking adequate irrigation water, are dry farmed to wheat and barley. With limited natural precipitation, a grain-fallow system is used. The increasing amounts of gravel and cobblestones with depth limits the available water holding capacity. Yields are low without irrigation. These soils are easy to cultivate because slopes are not steep and the surface soil is friable. Erosion is not a problem. Some areas of this soil contain cobbletones on the surface which can interfere with tillage.

Driggs Variant soils are good sources of gravel for road construction and other uses. The material in the C horizon has very rapid permeability which could pose problems of groundwater contamination if septic systems are used.

Soil 37

A—0 to 17 inches; very dark grayish brown silt loam, granular and subangular blocky structure, few gravel, neutral.

Btk—17 to 31 inches; dark grayish brown silty clay loam, subangular blocky structure, few gravel, slightly effervescent, mildly alkaline.

Bk1—31 to 39 inches; very pale brown silty clay loam, subangular blocky structure, few gravel, violently effervescent, moderately alkaline.

Bk2—39 to 56 inches; pale brown silty clay loam, subangular blocky structure, few gravel, violently effervescent, moderately alkaline.

Soil 37 is in the foreground with sagebrush. Cirques in the mountains contained small ice fields from which glaciers moved.

Soil 37

Family Classification: fine-loamy, mixed, Argic Pachic Cryoborolls

Soil Depth: 60 inches or more

Drainage Class: well drained

Parent Material: glacial outwash

Average Annual Precipitation: about 17 inches

Average Annual Air Temperature: about 37 degrees F.

Average Frost-Free Season: 50 days or less

Elevation: 7,000 to 8,000 feet

Topography: gently and moderately sloping terraces

Habitat Type: *Artemisia tridentata vaseyana spiciformis/Festuca idahoensis*

Common Native Vegetation: subalpine big sagebrush, Idaho fescue, mountain brome, bluegrass, sedge, arrowleaf balsamroot, tapertip hawksbeard, and lupine

Occurrence in Idaho: eastern part, mainly in Lemhi county

Land Use: rangeland

The Prominent Characteristics of This Soil Are: thick, very dark colored surface layer; lime accumulation deep in the soil; short frost-free season

Most of the higher mountains of Idaho had small alpine glaciers during the ice age. **Soil 37** formed in limestone material which was washed from nearby glacial moraines deposited by these glaciers. This soil occurs at the foot of limestone mountains in a somewhat dry cool climate.

Soil development, as is usual in cool climates, has been slow. Lime, present in the parent material, leached out of the surface soil and is now in the subsoil. This is similar to what occurred in **Soil 46 (Rexburg Series).**

Some movement of clay from the surface layer into the subsoil has also taken place. This is evident not only by more clay being present in the lower soil horizons, but also by the occurrence of clay films clinging to the surfaces of soil aggregates.

Sagebrush is common on this soil; however, the cool climate has been favorable for many grasses to flourish. This has created a thick dark surface soil similar to **Soil 49 (Ola Series).**

Soil 37 has had recent extensive mixing of the surface layers by large burrowing animals, possibly badgers. The wavy lower boundary of the A horizon and the presence of a krotovina in the upper right hand part of the soil profile photo shows this mixing. This genetic process is discussed briefly on page 9.

The short frost-free season and cool summer temperatures restricts the agricultural use of this soil to rangeland. With proper stocking rates and timing of grazing, this soil produces some of the highest yields of forage for soils in Idaho having a sagebrush habitat type.

Soil 38 (Greys Series)

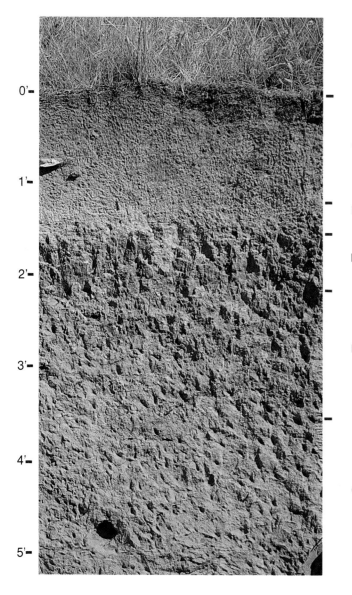

A—0 to 15 inches; grayish brown silt loam, granular and subangular blocky structure, neutral.

E—15 to 19 inches; pale brown silt loam, subangular blocky structure, neutral.

BE—19 to 25 inches; brown and pale brown silt loam, subangular blocky structure, neutral.

Bt—25 to 43 inches; brown silt loam, subangular blocky structure, neutral.

C—43 to 63 inches; light yellowish brown silt loam, massive, mildly alkaline.

Soil 38 (Greys Series) with hay and pasture. The dense stand of quaking aspen on this soil is a colorful native species. Mountains in the distance are in the Targhee National Forest.

Soil 38 (Greys Series)

Family Classification: fine-silty, mixed Boralfic Cryoborolls

Soil Depth: 60 inches or more

Drainage Class: well drained

Parent Material: loess

Average Annual Precipitation: about 18 inches

Average Annual Air Temperature: about 39 degrees F.

Average Frost-Free Season: about 70 days

Elevation: 5,600 to 7,000 feet

Topography: gently sloping to hilly uplands

Habitat Type: *Populus tremuloides/Calamagrostis rubescens*

Common Native Vegetation: quaking aspen, snowberry, common chokecherry, pine reedgrass, blue wildrye, mountain brome, slender wheatgrass, sedge, lupine, and geranium

Occurrence in Idaho: eastern part, mainly in Fremont and Teton counties

Land Use: woodland and nonirrigated cropland

The Prominent Characteristics of This Soil Are: E soil horizon below the surface soil; short frost-free season; susceptible to frost action

The Greys soils occur in areas having a cool, short growing season. Frosts in summer months occur in some years. This greatly limits the successful growing of cultivated crops. A few areas have been cleared, however, for wheat, barley, and oats. Yields are not only marginal but inconsistent. Crops are harvested late and winter sometimes intervenes. In addition, there is a high erosion hazard on moderately steep slopes.

The natural vegetation of aspen and shrubs is excellent habitat for wildlife. Big game, such as deer and elk, migrate from nearby mountainous areas in winter months to escape the deep snows and feed on the plentiful vegetation.

These soils provide good summer grazing for livestock. However, they tend to be overgrazed as cattle concentrate in these areas because of the shade provided by the aspen.

Some aspen is cut for firewood and fence posts although the soft wood is of low quality for these uses.

In the development of these soils the climate was moist enough for the lime from the original calcareous loess to be leached entirely from the soil, except in a few places. In those places, it is in the deep substratum. This leaching process also was responsible for the development of the E soil horizon. Not only has all the lime been removed from this horizon by intense leaching, but also much of the clay that was formed here.

This development contrasts with **Soil 46 (Rexburg Series),** which also formed in loess about a thousand feet lower in elevation. The drier and warmer climate of the Rexburg soils failed to leach the lime below the subsoil and the soils are alkaline throughout.

The Greys soils, with its silty textures, is subject to frost heaving during prolonged cold weather. Buildings and other structures must be designed to counter this soil behavior.

Soil 39

A—0 to 3 inches; grayish brown gravelly silt loam, platy structure, medium acid.

Bt1—3 to 14 inches; brown gravelly silty clay loam, subangular blocky structure, medium acid.

2Bt2b—14 to 27 inches; brown clay, prismatic structure, medium acid.

2Bqmb—27 to 34 inches; pink duripan.

Soil 39 with a cover of low sagebrush and grasses.

Soil 39

Family Classification: very fine, mixed Duric Cryoborolls

Soil Depth: 20 to 40 inches to duripan

Drainage Class: well drained

Parent Material: alluvium over rhyolitic welded tuff residuum

Average Annual Precipitation: about 16 inches

Average Annual Air Temperature: about 40 degrees F.

Average Frost-Free Season: 50 days or less

Elevation: about 6,300 feet

Topography: nearly level to moderately sloping plains

Habitat Type: *Artemisia arbuscula/Festuca idahoensis*

Common Native Vegetation: low sagebrush, Idaho fescue, Sandberg bluegrass, and bluebunch wheatgrass

Occurrence in Idaho: southwestern part, mainly in Owyhee and Twin Falls counties

Land Use: rangeland

The Prominent Characteristics of This Soil Are: strongly cemented duripan; well developed subsoil

This soil developed in material weathered from rhyolitic welded tuff which is similar in composition to granite but is composed of finer crystals. Soils developed in granite, such as **Soil 17 (Shellrock Series),** are mostly sandy, whereas soils developed in rhyolitic material have finer textures.

Soil 39 shows a complex sequence of 4 periods of soil development. Over many thousands of years a duripan developed, cemented only by silica. All the duripans described in this book, such as **Soil 4 (Chilcott Series)** and **Soil 5 (Colthorp Series),** are cemented by lime as well as silica. Both kinds of pans effectively limit root growth.

The second period of development was the formation of an A and B horizon sequence. Over a long period of time the B horizon became well developed, becoming one of the most clayey soils in this book.

The next period happened during the ice age. Although this soil was many miles from the glaciers, it remained frozen much of the time. Partial thawing resulted in uneven erosion and movement of soil, mixing the A horizon and the upper part of the B horizon. The upper 14 inches of the present soil is this material. This process is similar to the development of patterned ground described in **Soil 14 (Flybow Series).** The mounds where Soil 39 occurs are more subdued than the mounds shown with the Flybow Series.

The last period of development has been the formation of the present A horizon and the clay enriched Bt1 horizon in the upper 14 inches.

Soil 39 is suited agriculturally only for rangeland, mainly due to the cool short growing season. Even if temperatures were more favorable, the moderate depth to the duripan, as well as the clayey buried subsoil, would limit the available water holding capacity.

Engineering uses of this soil have to contend with the pan as well as the clayey buried subsoil which has high shrink-swell properties. Septic systems using this soil would not work.

Soil 40 (Pavohroo Series)

A1—0 to 4 inches; very dark gray silt loam, granular structure, slightly acid.

A2—4 to 22 inches; very dark grayish brown silt loam, blocky structure, few gravel, slightly acid.

Bw1—22 to 44 inches; dark grayish brown silt loam, blocky structure, few gravel, slightly acid.

Bw2—44 to 55 inches; brown gravelly loam, massive, mildly alkaline.

C—55 to 63 inches; pale brown very cobbly loam, massive, slightly effervescent, mildly alkaline.

Soil 40 (Pavohroo Series) is in the background under Douglas-fir. Soils in the foreground have brown colored A horizons.

Soil 40 (Pavohroo Series)

Family Classification: fine-loamy, mixed pachic Cryoborolls
Soil Depth: 60 inches or more
Drainage Class: well drained
Parent Material: loess over limestone residuum
Average Annual Precipitation: about 24 inches
Average Annual Air Temperature: about 40 degrees F.
Average Frost-Free Season: 50 days or less
Elevation: 5,600 to 7,400 feet
Topography: moderately steep and steep foothills and mountains
Habitat Type: *Psuedotsuga menziesii/Calamagrostis rubescens*
Common Native Vegetation: Douglas-fir, low Oregon-grape, snowberry, pine reedgrass, sedge, blue wildrye, bearded wheatgrass, and geranium
Occurrence in Idaho: southeastern part, mainly from Bingham to Bear Lake counties
Land Use: woodland and watershed

The Prominent Characteristics of This Soil Are: thick, very dark-colored surface layer; short frost-free season

The Pavohroo soils would, by appearance, seem to be ideal garden soils. They have adequate depth for good moisture holding capacities with no layers to impede root development. The clay content is about 15 to 20 percent throughout, which is very good for holding plant nutrients. The organic matter is high and continues deep into the soil, which is evidenced by the dark colors. Best of all they could be cultivated with ease.

The thing that cannot be seen in the soils, however, is the cool temperatures with the possibility of frosts any time of the year. This, along with the steep slopes, is quite effective in limiting the use of these soils to woodland.

The habitat is excellent for wildlife, including mule deer and elk.

Even though the Pavohroo soils developed in parent material containing calcium carbonate, the climate has been moist enough to leach the lime almost entirely from the soils. In contrast, other soils which developed in limey materials in cool, but drier climates, like **Soil 37**, leached the lime to shallower depths.

Soil 1, occurring in the same general area as Pavohroo soils and having the same kind of parent material with comparable topography, shows a strikingly different kind of soil development. The average elevation is a few hundred feet higher and the temperature a bit cooler so that alpine fir grows. But this is not enough to account for such a difference. The extensive leaching and the movement of clay into the well developed subsoil of **Soil 1** shows the effect of much greater age. Pavohroo soils are comparatively young.

Soil 41 (Tannahill Series)

A—0 to 11 inches; dark brown extremely gravelly loam, granular structure, mildly alkaline.

Bt—11 to 23 inches; brown extremely gravelly loam, subangular blocky structure, mildly alkaline.

Bk1—23 to 32 inches; brown extremely gravelly loam, subangular blocky structure, common pockets and veins of lime, mildly alkaline.

Bk2—32 to 45 inches; very pale brown very gravelly loam, subangular blocky structure, strongly effervescent, strongly alkaline.

R—45 to 52 inches; basalt bedrock.

Soil 41 (Tannahill Series) associated with many basalt cliffs.

Soil 41 (Tannahill Series)

Family Classification: loamy-skeletal, mixed, mesic Calcic Argixerolls

Soil Depth: 40 to 60 inches or more to basalt bedrock

Drainage Class: well drained

Parent Material: loess mixed with basalt colluvium

Average Annual Precipitation: about 14 inches

Average Annual Air Temperature: about 52 degrees F.

Average Frost-Free Season: about 170 days

Elevation: 800 to 2,800 feet

Topography: steep and very steep canyons

Habitat Type: *Agropyron spicatum/Poa sandbergii*

Common Native Vegetation: bluebunch wheatgrass, Sandberg bluegrass, arrowleaf balsamroot, eriogonum, biscuitroot, and lupine

Occurrence in Idaho: north-central part, mainly in Idaho county

Land Use: rangeland

The Prominent Characteristics of This Soil Are: more than 35 percent coarse fragments; lime accumulation deep in the soil; steep slopes

Tannahill soils occur on slopes of deep canyons along the Salmon and Snake Rivers and their tributaries. Long steep slopes with gradients up to 90 percent are common. Rock outcrop is also common. This is a very scenic part of Idaho where Hells Canyon of the Snake River, the deepest gorge on earth occurs. Great changes in climate and natural vegetation occur in these canyons, either with abrupt changes in elevation or changes from north-facing to south-facing slopes.

Tannahill soils are at the lower elevations with south-facing exposures, so they are the warmest and driest soils in these canyons. They have a longer frost-free season each year than any of the soils in this book. These soils are intermediate between dry soils, like **Soil 4 (Chilcott Series)**, and moist soils, like **Soil 50 (Palouse Series)**. They were dry enough so that lime was not leached entirely out of the profile but were moist enough to have a good grass cover and form a dark colored surface soil.

Surprisingly, the landscape has been stable long enough so that these soils had time to form Bt horizons. Even though there has been high runoff on these steep slopes, there has been enough gravel and cobblestones on the surface to slow the natural erosion to allow for soil development to take place. The moderate amount of rainfall and warm temperatures allowed weathering and movement of clay into the subsoil.

These soils are used almost entirely for rangeland. When the range is in good condition, forage yields are good. If the range is in poor condition, reseeding may be done on the more moderately sloping areas.

The steep slopes is the overriding limitation to most other uses. Recreational development has potential. Careful placement of paths and trails on these soils would take advantage of the scenic quality of the area.

Soil 42 (Gem Series)

A—0 to 7 inches; dark brown clay loam, granular structure, hard when dry, few gravel, slightly acid.

Bt—7 to 22 inches; dark brown clay, prismatic and subangular blocky structure, very hard when dry, few gravel, neutral.

R—22 to 35 inches; basalt bedrock, few cracks in upper part filled with brown slightly effervescent soil material.

Soil 42 (Gem Series) with natural vegetation of sagebrush and grasses.

Soil 42 (Gem Series)

Family Classification: fine, montmorillonitic, mesic Calcic Argixerolls

Soil Depth: 20 to 40 inches to basalt bedrock

Drainage Class: well drained

Parent Material: basalt residuum

Average Annual Precipitation: about 14 inches

Average Annual Air Temperature: about 47 degrees F.

Average Frost-Free Season: about 125 days

Elevation: 2,600 to 5,000 feet

Topography: gently sloping to steep uplands

Habitat Type: *Artemisia tridentata xericensis/Agropyron spicatum*

Common Native Vegetation: xeric big sagebrush, bluebunch wheatgrass, Idaho fescue, Sandberg bluegrass, arrowleaf balsamroot, and lupine

Occurrence in Idaho: southwestern part, mainly in Gem and Payette counties

Land Use: rangeland and nonirrigated and irrigated cropland

The Prominent Characteristics of This Soil Are: well developed subsoil; clayey subsoil with high shrink-swell; moderately deep to bedrock

Where the Gem soils are used for rangeland, it usually is in poor condition as a result of many years of overgrazing. Annual forage production is very low. Undesirable species such as medusahead wildrye, cheatgrass, and xeric big sagebrush predominate. Forage yields can be increased by careful management of grazing.

Some areas of these soils are dry farmed to small grains, hay and pasture. The annual precipitation is not sufficient however for successful annual cropping. Yields are poor because of the lack of available moisture during the growing season. The restricted depth of this soil is less than ideal for both root development and moisture holding capacity.

Areas of these soils that have a dependable supply of water are being irrigated and produce pasture, hay, and small grains. Careful design of the irrigation system is necessary to allow water to move through the clayey subsoil. Permeability is slow.

The Gem soils are somewhat similar to **Soil 52 (Magic Series).** They developed from the same basalt parent material and basalt bedrock occurs at the similar depths. The Magic soil contains more clay and has vertical cracks that extend to the surface resulting in a churning action which allows soil granules at the surface to fall into the cracks. The soil is continually being mixed to the depth of the cracking. The Gem soils are almost to that stage. They have a well-developed prismatic structure in the subsoil with vertical cracks extending upwards. It already contains enough expanding clay type to have a high shrink-swell potential. All that is required is for a bit more weathering to take place over possibly a few thousand years and the Gem soils will become the Magic soils!

Soil 43 (Gwin Series)

A—0 to 3 inches; very dark grayish brown gravelly silt loam, granular structure, neutral.

Bt1—3 to 10 inches; very dark grayish brown very gravelly clay loam, angular blocky structure, neutral.

Bt2—10 to 20 inches; dark yellowish brown very cobbly clay loam, prismatic structure, neutral.

R—20 to 36 inches; basalt bedrock.

Soil 43 (Gwin Series) with rangeland. Rock outcrop and moderately deep soils are closely associated. Very deep soils supporting trees are on north-facing slopes.

Soil 43 (Gwin Series)

Family Classification: loamy-skeletal, mixed, mesic Lithic Typic Argixerolls

Soil Depth: 10 to 20 inches to basalt bedrock

Drainage Class: well drained

Parent Material: loess mixed with basalt colluvium

Average Annual Precipitation: about 20 inches

Average Annual Air Temperature: about 48 degrees F.

Average Frost-Free Season: about 140 days

Elevation: 2,500 to 4,300 feet

Topography: moderately sloping to steep canyons

Habitat Type: *Festuca idahoensis/Agropyron spicatum*

Common Native Vegetation: Idaho fescue, bluebunch wheatgrass, Sandberg bluegrass, arrowleaf balsamroot, penstemon, and tapertip hawksbeard

Occurrence in Idaho: north-central part, from Idaho to Latah counties

Land Use: rangeland

The Prominent Characteristics of This Soil Are: shallow to bedrock; well developed subsoil; more than 35 percent coarse fragments

The Gwin soils, even though occurring on mostly south-facing steep canyon slopes, have well developed Bt horizons. These steep slopes have been stable for a long time in a climate that was moist enough and warm enough for active weathering of the loess parent material. The Bt horizons contain 27 to 35 percent clay. While thick deposits of loess were accumulating on nearby plateaus, most of the loess falling on the canyon slopes was being quickly eroded. Therefore the Gwin soils are shallow.

The steep slopes and many surface rock fragments limit the agricultural use of these soils to rangeland. Forage production, even if the range is in good condition, is low due to the total available water holding capacity of only 1½ to 3 inches. The very low available water holding capacity results not only from the shallowness of the soils but also from the high percentage of gravel, cobblestones, and stones.

All other uses of these soils are greatly restricted. Sanitary facilities are, for all practical purposes, impossible to install. Building site development is costly, mainly due to the shallow depth to bedrock.

Gwin soils are similar to **Soil 48 (Hymas Series)** in slope, high percentage of rock fragments, and the shallowness to bedrock. So the main limitations for use are similar. The main contrast in these soils is the limestone parent material of the Hymas soils. Before B horizons, enriched by clay, can develop, calcium carbonate must be leached from the soils.

Soil 44 (Klickson Series)

A—0 to 8 inches; dark grayish brown very cobbly silt loam, granular structure, slightly acid.

Bw—8 to 18 inches; brown very gravelly silt loam, subangular blocky structure, slightly acid.

Bt1—18 to 33 inches; yellowish brown very gravelly silt loam, subangular blocky structure, slightly acid.

Bt2—33 to 62 inches; yellowish brown very cobbly loam, subangular blocky structure, slightly acid.

Soil 44 (Klickson Series) is in woodland with steep slopes. Grass covered south-facing slopes in the distance are shallow to bedrock.

Soil 44 (Klickson Series)

Family Classification: loamy-skeletal, mixed, frigid Ultic Argixerolls

Soil Depth: 60 inches or more

Drainage Class: well drained

Parent Material: loess mixed with basalt colluvium

Average Annual Precipitation: about 26 inches

Average Annual Air Temperature: about 42 degrees F.

Average Frost-Free Season: about 80 days

Elevation: 2,000 to 5,000 feet

Topography: moderately steep to very steep canyons

Habitat Type: *Pseudotsuga menziesii/Physocarpus malvaceous*

Common Native Vegetation: Douglas-fir, ponderosa pine, mallow ninebark, creambush oceanspray, snowberry, rose, sedge, Columbia brome, white spirea, and heartleaf arnica

Occurrence in Idaho: northern part, mainly from Idaho to Latah counties

Land Use: woodland, hay, and pasture

The Prominent Characteristics of This Soil Are: more than 35 percent coarse fragments; steep slopes

Klickson soils developed on slopes of canyons. In the slow process of deepening these canyons, there has been an active downslope movement of loosened basalt fragments from the bedrock. Since these pieces of rock moved only relatively short distances they remained angular or sharp edged. This is in contrast to gravel and cobblestones rounded by moving longer distances by ice or water. An example of rounded gravel is evident in **Soil 45 (Little Wood Series).**

At the same time of the colluvial activity which took place on the canyon slopes, a moderate amount of loess was added to the debris. This silt loam textured loess, which has a high moisture holding capacity, improved the available moisture holding capacity of this soil considerably. Although the total amount is about 6 to 8 inches, it would have been much lower without the addition of loess. In contrast, **Soil 25** lacks loess in its profile. To the extent that coarse fragments reduce the total available water capacity, they do little harm by their presence when used for forestry.

Most areas of the Klickson soils are used for forestry. Production is good. Water movement and root development are unimpeded. The main problem in woodland management is with slopes steeper than about 35 percent. There is not only a limitation to the equipment used but there is a severe erosion hazard in disturbed areas.

Some areas, with moderately steep slopes, have been cleared and used for hay and pasture. Also in these slopes individual homesites have been developed. The large amount of rock fragments is a problem and properly designed sanitary facilities are difficult to build.

These soils also have limited use for grazing, especially if the tree canopy is opened by fire or logging.

Soil 45 (Little Wood Series)

Ap—0 to 11 inches; dark brown gravelly loam, granular structure, slightly acid.

Bt—11 to 22 inches; dark yellowish brown very gravelly sandy clay loam, subangular blocky structure, medium acid.

BC—22 to 32 inches; brown extremely gravelly sandy loam, subangular blocky structure, slightly acid.

C1—32 to 47 inches; dark yellowish brown extremely gravelly loamy sand, single grain, slightly acid.

C2—47 to 60 inches; yellowish brown extremely gravelly coarse sand, single grain, neutral.

Soil 45 (Little Wood Series) with rye in the foreground and pasture across the road. Moderately deep sandy soils are in rangeland with steep slopes. Very deep sandy soils are at higher elevations under trees.

Soil 45 (Little Wood Series)

Family Classification: loamy-skeletal, mixed, frigid Ultic Argixerolls

Soil Depth: 60 inches or more

Drainage Class: well drained

Parent Material: alluvium

Average Annual Precipitation: about 14 inches

Average Annual Air Temperature: about 43 degrees F.

Average Frost-Free Season: about 85 days

Elevation: 4,700 to 5,500 feet

Topography: nearly level and gently sloping terraces

Habitat Type: *Artemisia tridentata vaseyana/Festuca idahoensis*

Common Native Vegetation: mountain big sagebrush, antelope bitterbrush, Idaho fescue, bluebunch wheatgrass, western needlegrass, Sandberg bluegrass, and arrowleaf balsamroot

Occurrence in Idaho: southern part, mainly in Blaine and Camas counties

Land Use: rangeland and nonirrigated and irrigated cropland

The Prominent Characteristics of This Soil Are: more than 35 percent coarse fragments; gravel and sand substratum

The Little Wood soils occur at the foot of mountains on stream terraces. During more rainy climates of the past, repeated floods carrying gravel and even cobblestones came out of the mountain valleys. As the gradient of the streams became gentle, the huge amount of coarse fragments being carried along by the water, settled out onto broad nearly level areas. The gravel and cobblestones are rounded showing the effect of being moved long distances by swiftly-flowing water. **Soil 25,** in contrast, has angular coarse fragments, which were moved only a short distance down steep and very steep slopes.

As time passed, the climate of the area in which Little Wood soils developed became drier with no large-scale flooding. The topography stabilized. This allowed the slow processes of weathering to take place. The Bt soil horizon received clay which moved downward from the surface soil.

The dark color of the surface soil shows that organic matter accumulated in this cool prairie environment.

The low total available water holding capacity of 3 to 6 inches lowers expected yields when these soils are used for dryland agriculture. Yields improve significantly with irrigation when alfalfa hay, pasture, and small grains are grown.

Forage yields on rangeland are only moderate because of the droughty soils even when the range is in good condition.

These soils are excellent sources of gravel for construction of roads and other uses. In places there are subdivision developments taking place. The main concern is the possible pollution of groundwater caused by the excessive permeability of the substratum.

Soil 46 (Rexburg Series)

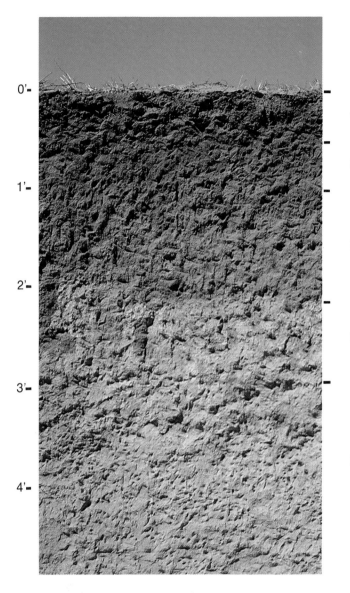

Ap—0 to 6 inches; grayish brown silt loam, granular structure, mildly alkaline.

Bw1—6 to 12 inches; brown silt loam, subangular blocky structure, mildly alkaline.

Bw2—12 to 25 inches; pale brown silt loam, subangular blocky structure, mildly alkaline.

Bk1—25 to 35 inches; white silt loam, subangular blocky structure, violently effervescent, moderately alkaline.

Bk2—35 to 59 inches; light gray silt loam, massive, violently effervescent, moderately alkaline.

Soil 46 (Rexburg Series) with a crop of irrigated potatoes in the foregound and winter wheat in the distance.

Soil 46 (Rexburg Series)

Family Classification: coarse-silty, mixed, frigid Calcic Haploxerolls

Soil Depth: 60 inches or more

Drainage Class: well drained

Parent Material: loess

Average Annual Precipitation: about 14 inches

Average Annual Air Temperature: about 43 degrees F.

Average Frost-Free Season: about 90 days

Elevation: 4,800 to 6,000 feet

Topography: gently sloping to steep uplands

Habitat Type: *Artemisia tripartita/Agropyron spicatum*

Common Native Vegetation: threetip sagebrush, antelope bitterbrush, bluebunch wheatgrass, Idaho fescue, Nevada bluegrass, and arrowleaf balsamroot

Occurrence in Idaho: southeastern part, mainly from Fremont to Power counties

Land Use: rangeland and nonirrigated and irrigated cropland

The Prominent Characteristics of This Soil Are: lime accumulation deep in the soil; weakly developed subsoil; moderate permeability

Rexburg soils are considered to be one of the better dry farmed soils in southeastern Idaho. This part of the state has somewhat limited precipitation so that summer fallow is common. These soils, formed in deep loess, have a high available water capacity so that moisture can be stored for the following years crops. Yields are moderately good. Tillage is easy but soil erosion, both by water and wind, can be a problem in places.

Where water is available, these soils are irrigated. Most areas of these soils are sloping so sprinkler irrigation is often used. Yields are very good. With irrigation, potatoes are an important crop.

The loess, from which Rexburg soils developed, was calcareous when first laid down by the wind. Most of the loess came from the nearby Snake River plain. Floods from glacial meltwater provided the source of most of this material.

These soils developed in a cool relatively dry climate. The climate was moist enough to leach the lime to lower depths in the soils, but not enough to leach it completely as in **Soil 50 (Palouse Series).** Weathering in the Rexburg soils also took place at a slower rate than the somewhat similar Palouse soils because of the cooler drier climate. Textures of both soils are silt loam but the percentage of clay in the Palouse soils is nearly double that of the Rexburg soils. The structure of the Rexburg soils also is not as strongly developed. Notice the darker colors of the Palouse soils. Grasses in this more moist climate provided more organic matter to the soils.

This is an excellent example showing the effects of different climates on a similar parent material.

Septic tank filter fields have the greatest chance for success on the gently sloping areas of the Rexburg soils of any other soil in this book. It is not only a very deep soil but the permeability of the subsoil is just about right — neither too fast nor too slow. This allows for complete bacterial activity to take place without contamination of the deeply lying water table.

Soil 47 (Westlake Series)

A—0 to 4 inches; black silt loam, granular structure, neutral.

Ag—4 to 28 inches; black silt loam, mottles, subangular blocky structure, neutral.

Bg—28 to 52 inches; very dark gray silty clay loam, mottles, prismatic structure, neutral.

Cg—52 to 66 inches; very dark grayish brown silty clay loam, mottles, massive, few coarse fragments, ground water seeping out of lower part, neutral.

Colors are for moist soil.

Soil 47 (Westlake Series) in pasture along the stream. Shallow, stony soils are on the left and very deep, well drained soils are on the right.

Soil 47 (Westlake Series)

Family Classification: fine-silty, mixed, frigid Cumulic Ultic Haploxerolls

Soil Depth: 60 inches or more

Drainage Class: somewhat poorly drained

Parent Material: alluvium

Average Annual Precipitation: about 22 inches

Average Annual Air Temperature: about 44 degrees F.

Average Frost-Free Season: about 100 days

Elevation: 3,000 to 4,000 feet

Topography: nearly level bottomlands

Habitat Type: unclassified

Common Native Vegetation: willow, redosier dogwood, tufted hairgrass, sedge, Baltic rush, blue wildrye, common camas, common cowparsnip, and biscuitroot

Occurrence in Idaho: north-central part, mainly from Idaho to Latah counties

Land Use: hay and pasture

The Prominent Characteristics of This Soil Are: thick, very dark colored surface layer; fluctuating water table; flooding hazard

The photo of Westlake soils looks somewhat like **Soil 18 (Pywell Series)** in being very dark colored throughout. Pywell soils, however, consist of very deep peat and their physical properties are totally different.

Westlake soils are more like **Soil 34 (Houk Series).** These mineral soils both formed in alluvium, they occur in nearly level bottomland positions, have high seasonal water tables, and are subject to flooding. So, these soils have similar problems when used for agriculture or building development.

There are, however, some differences that may affect their use and management. Westlake soils lack the sandy layers which Houk soils have in the lower part. Westlake soils are also free of lime. These soils occur in areas having more precipitation and warmer temperatures than Houk soils. Ranker growth of grasses has produced more organic matter throughout the profile making them darker colored. Their fertilizer needs are different.

Although Westlake soils are used mostly for hay and pasture, there are places where drainage has been improved and wheat and barley are grown. Yields are good.

Construction sites for buildings and roads must contend with the high water table and the possibility of flooding. Too often this is not recognized during the dry part of the year and serious errors in planning are made. These errors can be difficult and expensive to correct. It would be better to look elsewhere for these uses.

Soil 48 (Hymas Series)

A—0 to 4 inches; dark grayish brown cobbly silt loam, granular structure, slightly effervescent, neutral.

Bw—4 to 8 inches; brown cobbly silt loam, blocky structure, strongly effervescent, mildly alkaline.

Bk—8 to 19 inches; white very stony silt loam, massive, violently effervescent, mildly alkaline.

R—19 to 25 inches; limestone bedrock.

Soil 48 (Hymas Series) is on the ridges. Similar but deeper soils are on the slopes.

Soil 48 (Hymas Series)

Family Classification: loamy-skeletal, carbonatic, frigid Lithic Haploxerolls
Soil Depth: 10 to 20 inches to limestone bedrock
Drainage Class: well drained
Parent Material: limestone residuum
Average Annual Precipitation: about 13 inches
Average Annual Air Temperature: about 43 degrees F.
Average Frost-Free Season: about 80 days
Elevation: 4,600 to 7,000 feet
Topography: moderately sloping to steep ridges of hills and mountains
Habitat Type: *Artemisia nova/Agropyron spicatum*
Common Native Vegetation: black sagebrush, bluebunch wheatgrass, Sandberg bluegrass, and lupine
Occurrence in Idaho: southern part, mainly in Bingham to Cassia counties
Land Use: rangeland

The Prominent Characteristics of This Soil Are: weakly developed subsoil; shallow to bedrock

The Hymas soils are similar to **Soil 43 (Gwin Series),** in being shallow to bedrock, in having a large amount of coarse fragments, and in having moderately sloping to steep slopes. These properties give these soils similar limitations in use. Hymas soils formed from parent material weathered from limestone, whereas Gwin soils developed from loess mixed with basalt colluvium.

Note that there is still lime in the surface soil of the Hymas profile. The effervescence of the soil when treated with diluted acid shows the presence of lime. Lime must be leached from the soil before Bt horizons can begin to develop. With the present limited precipitation and with the soils being on ridges with high runoff, this seems unlikely to happen for a very long time.

These soils have limited use. They are mostly for rangeland. Yields are low and it is difficult to maintain the range in good condition. Reseeding is hampered by the coarse fragments in the surface soil.

Limited use is made of these soils by wildlife. Nearby deeper soils have much better habitat for wildlife. The ridge positions that these soils occupy are ideal areas for paths and trails.

Little use is made of these soils for construction sites because of inaccessibility and the shallow depth to bedrock.

Soil 49 (Ola Series)

A1—0 to 16 inches; very dark grayish brown loam, granular structure, few gravel, slightly acid.

A2—16 to 29 inches; dark grayish brown gravelly sandy loam, massive, slightly acid.

Cr—29 to 43 inches; weathered gneiss.

Soil 49 (Ola Series) with a mixture of shrubs and grasses.

Soil 49 (Ola Series)

Family Classification: coarse-loamy, mixed, frigid Pachic Haploxerolls

Soil Depth: 20 to 40 inches to gneiss bedrock

Drainage Class: well drained

Parent Material: gneiss residuum

Average Annual Precipitation: about 18 inches

Average Annual Air Temperature: about 44 degrees F.

Average Frost-Free Season: about 85 days

Elevation: 3,500 to 4,500 feet

Topography: moderately steep to very steep hills and mountains

Habitat Type: *Artemisia tridentata vaseyana/Festuca idahoensis*

Common Native Vegetation: mountain big sagebrush, blue elderberry, Idaho fescue, bluebunch wheatgrass, Sandberg bluegrass, and arrowleaf balsamroot

Occurrence in Idaho: southwestern part, mainly in Ada and Gem counties

Land Use: rangeland

The Prominent Characteristics of This Soil Are: thick, very dark colored surface layer; no subsoil development; easily rippable bedrock

Ola soils are similar, in a way, to many other soils which occur only on steeply-sloping areas. Because of the steep slopes, they all have great limitations. Uses of these soils, other than utilizing the existing natural vegetation, is very limited. Even then, great care must be used in their management or siltation from erosion is likely to take place on lower-lying soils and in drainageways.

Constructing roads on these soils must be done with great care. These soils are fragile and erode easily. This is also true with other kinds of building development. The steep slope is the greatest limiting factor.

Ola soils formed in residuum from gneiss rock, mostly on north-facing slopes. Thick A horizons developed in this environment. Not only was there an abundance of grasses to provide organic matter, but the steep north-facing slopes were cool which slowed the destruction of organic matter. On nearby south-facing slopes which are much warmer, destruction of organic matter is more complete. This contrast of thickness of A horizons between north and south-facing slopes is common at these latitudes.

Ola soils did not develop B horizons rich in clay. This is common with soils having not only cool climates but having steep slopes. It takes far more time to develop these kinds of B horizons in cool moist climates than with soils having warm more moist climates. Also the natural geologic erosion on steep slopes continues at about the same rate as does the weathering of the gneiss bedrock to parent material. This keeps the soils young, allowing only A horizons to develop. Another soil with a similar lack of B horizon development is **Soil 17 (Shellrock Series).**

Soil 50 (Palouse Series)

Ap—0 to 8 inches; dark grayish brown silt loam, granular structure, slightly acid.

Bw—8 to 27 inches; dark brown silt loam, subangular blocky structure, neutral.

Bt1—27 to 41 inches; yellowish brown silt loam, uncoated silt grains on prismatic and subangular blocky structure, neutral.

Bt2—41 to 63 inches; yellowish brown silty clay loam, prismatic and subangular blocky structure, neutral.

Soil 50 (Palouse Series) with a field of lentils. North-facing soils in the background have E horizons below the surface layer.

Soil 50 (Palouse Series)

Family Classification: fine-silty, mixed, mesic Pachic Ultic Haploxerolls
Soil Depth: 60 inches or more
Drainage Class: well drained
Parent Material: loess
Average Annual Precipitation: about 21 inches
Average Annual Air Temperature: about 48 degrees F.
Average Frost-Free Season: about 140 days
Elevation: about 2,700 feet
Topography: gently sloping to hilly uplands
Habitat Type: *Festuca idahoensis/Symphoricarpos albus*
Common Native Vegetation: Idaho fescue, bluebunch wheatgrass, snowberry, rose, and white spirea
Occurrence in Idaho: northern part, mainly in Benewah, Latah, and Nez Perce counties
Land Use: nonirrigated cropland

The Prominent Characteristics of This Soil Are: high available water holding capacity; thick dark colored surface layer; very deep soil; susceptible to frost action

The Palouse soils occur in the area of northern Idaho which received the most loess. Loess depths of at least 100 feet are common. Deposits resulted from strong winds picking up mostly silt size particles from south-central Washington and carrying them eastward into Idaho.

Continental glaciers advanced at least 7 times into what is now the United States from Canada. Outwash from the melting glaciers provided a huge supply of silty material which was the major source of the loess. The Palouse soils developed in only the latest major deposit. This contrasts with **Soil 35 (Nez Perce Series)** which developed in all the deposits and **Soil 33 (Southwick Series)** which developed in the last two deposits.

The natural high fertility and consistent production of the Palouse soils were quickly discovered by the early settlers. Crop failures due to drought are very rare. This is due mostly to the climate and the kinds of crops grown. Most of the precipitation comes during winter and early spring months, filling the soil adequately for high yields of winter wheat. Spring seeded crops of wheat, barley, dry peas, and lentils require a few timely spring rains to produce maximum yields.

Spring and summer soil temperatures are also nearly ideal for all of these crops. Summer rainfall is low so crops that require summer moisture are not grown. There is no large supply of water available for general irrigation and the hilly topography would make it difficult.

Palouse soils, in addition to being very deep, have a high available water holding capacity — 11 to 12 inches. Also, there are no subsoil layers which inhibit the movement of water or penetration of roots.

So for dryland cropping, everything seems to be ideal. It may be except for one thing — erosion. Control of soil erosion has been difficult to achieve. The steeper slopes are very susceptible, especially since much of the erosion is caused by water from melting snow. Winter wheat, seeded late enough to escape the threat of disease, does not put on enough growth to guard against high runoff.

Much of the original topsoil has been lost through erosion during the past century. Not only has the once excellent tilth been lost but much fertility. High amounts of fertilizer are now required.

Palouse soils, which are high in silt, have a high potential for frost action. This affects the construction of buildings, roads, and streets.

Soil 51

E—0 to 2 inches; dark brown silt loam, platy structure, extremely acid.

Bs—2 to 9 inches; yellowish red silt loam, granular structure, very strongly acid.

Bw—9 to 15 inches; dark reddish brown silt loam, granular structure, strongly acid.

2Ab—15 to 25 inches; dark brown silt loam, mottles, subangular blocky structure, few cobblestones, medium acid.

2Bwb—25 to 30 inches; dark yellowish brown silt loam, mottles, massive, strongly acid.

Colors are for moist soil.

Soil 51 is next to the lake. Soils on steep mountains are well drained and lack reddish colored subsoils.

Soil 51

Family Classification: coarse-loamy, mixed Entic Cryorthods
Soil Depth: 60 inches or more
Drainage Class: moderately well drained
Parent Material: volcanic ash over glacial till
Average Annual Precipitation: about 60 inches
Average Annual Air Temperature: about 36 degrees F.
Average Frost-Free Season: 50 days or less
Elevation: 5,000 to 6,600 feet
Topography: gently sloping to moderately steep cirque basins
Habitat Type: *Abies lasiocarpa/mensiesia ferruginea*
Common Native Vegetation: subalpine fir, Engelmann spruce, mock azalea, myrtle pachystima, big blueberry, and whiteflower rhododendron
Occurrence in Idaho: northern part, mainly from Boundary to Idaho counties
Land Use: woodland and watershed

The Prominent Characteristics of This Soil Are: highly erodible volcanic ash surface soil; reddish colored subsoil

This soil occurs in cirque basins throughout northern Idaho where snow depths of 10 feet or more are common by late winter. Snow in these areas often remains until summer. The large amount of water available for leaching the soil, along with cool temperatures, has resulted in a special kind of soil-forming process.

The dominant soil development processes of this young soil have been the dissolving of primary minerals and the movement of sesquioxides (compounds of iron and aluminum) and organic matter mainly into the Bs soil horizon, with lesser amounts into the Bw horizon. Iron oxides give the soil its red color. A very thin layer of darker colored soil can be seen between the E and Bs horizons where organic matter has accumulated.

Other soils in northern Idaho, such as **Soil 19 (Vay Series), Soil 21,** and **Soil 24,** are somewhat similar but have not had as much leaching as **Soil 51.** These three soils accumulated sufficient iron to give them colorful surface soils but not enough to make them as red as **Soil 51.**

This iron accumulation has taken place only in the volcanic ash layer in the upper part of the soil. This ash was deposited upon glacial till about 6,700 years ago, a result of the eruption of Mt. Mazama in southwestern Oregon.

The lower two soil horizons are actually the remains of an old soil which had developed before the addition of volcanic ash. Since the climate was much drier at that time, it formed a dark-colored A horizon with a moderate amount of organic matter, rather than the reddish colored layers which are now in the volcanic ash.

Soil 52 (Magic Series)

A—0 to 3 inches; brown very stony silty clay, granular structure, neutral.

Bw1—3 to 15 inches; brown silty clay, prismatic structure, wide cracks, few slickensides, mildly alkaline.

Bw2—15 to 29 inches; pinkish gray silty clay, prismatic structure, wide cracks, many slickensides, moderately alkaline.

Bk—29 to 39 inches; pinkish gray silty clay, angular blocky structure, slightly effervescent, moderately alkaline.

R—39 to 44 inches; basalt bedrock with calcium carbonate.

Soil 52 (Magic Series) in rabbitbrush. The Pioneer mountains near Hailey are in the far background.

Soil 52 (Magic Series)

Family Classification: fine, montmorillonitic, frigid Entic Chromoxererts
Soil Depth: 20 to 40 inches to basalt bedrock
Drainage Class: well drained
Parent Material: basalt residuum
Average Annual Precipitation: about 14 inches
Average Annual Air Temperature: about 41 degrees F.
Average Frost-Free Season: about 85 days
Elevation: 4,900 to 5,500 feet
Topography: nearly level and gently sloping plains
Habitat Type: *Artemisia longiloba/Festuca idahoensis*
Common Native Vegetation: alkali sagebrush, rabbitbrush, Idaho fescue, Sandberg bluegrass, western needlegrass, bottlebrush squirreltail, narrowleaf pussytoes, and hood phlox
Occurrence in Idaho: southern part, mainly Camas county
Land Use: rangeland and nonirrigated and irrigated cropland

The Prominent Characteristics of This Soil Are: clayey soil; wide cracks to the surface when dry; stones on the surface; very slow permeability

This soil has had a similar history of development as **Soil 53 (Ager Series)** and **Soil 54 (Boulder Lake Series).** Although the parent materials are different, they weathered over a very long period of time.

As clay was first formed in the basalt residuum, it was moved slowly by water into lower depths. When a moderately high amount of clay was attained, another genetic process started. The high shrink-swell property of the particular kind of clay formed resulted in vertical cracks appearing at the surface during prolonged dry periods. This allowed small quantities of soil granules on the surface to fall into the cracks, producing a churning action to the depth of cracking. The falling of granules into the cracks would have been caused by natural sloughing or any disturbance at the surface such as wind or animal movement. The formation of clay and churning of the soil by wetting and drying was faster than the natural soil erosion which took place. Therefore the soil became more and more clayey throughout, even in the surface layer, making the soil more resistant to erosion.

The churning action plucked basalt stones from the underlying fractured bedrock and lifted them to the surface. There are rarely any stones between the surface and the bedrock. This is a rather unusual situation. This can be an advantage if this soil is being prepared for cultivation. Unlike most other very stony soils, once the surface stones are removed, there is little likelihood of having to re-do the job periodically.

Cultivating the Magic soils, even after stone removal, is difficult because of the high clay content of the surface soil. There is always the hazard of plant roots being torn apart by the churning action.

In comparing the photograph of this soil with the Ager soil profile, notice that the Magic soil is a bit darker, especially in the upper parts. The cooler temperatures at the higher elevations, along with a little more precipitation, provided more vegetative growth, with a resultant higher organic matter content.

The moderate depth to bedrock, the high shrink-swell, and the very slow permeability all greatly limit the use of this soil for pipelines, highways, buildings, and septic systems.

Soil 53 (Ager Series)

A—0 to 4 inches; pale brown clay, granular structure, neutral.

Bw—4 to 35 inches; brownish yellow clay, prismatic structure, wide cracks, common slickensides, moderately alkaline.

Bk—35 to 48 inches; white clay, prismatic structure, wide cracks, slightly effervescent, moderately alkaline.

C—48 to 54 inches; very pale brown clay, massive, slightly alkaline.

Soil 53 (Ager Series) is in range on the hills. Soils in the foreground are in very deep silty alluvium.

Soil 53 (Ager Series)

Family Classification: very fine, montmorillonitic, mesic Entic Chromoxererts

Soil Depth: 60 inches or more

Drainage Class: well drained

Parent Material: tuff or siltstone residuum

Average Annual Precipitation: about 12 inches

Average Annual Air Temperature: about 48 degrees F.

Average Frost-Free Season: about 130 days

Elevation: 2,500 to 3,100 feet

Topography: gently sloping to steep uplands

Habitat Type: *Artemisia tridentata wyomingensis/Agropyron spicatum*

Common Native Vegetation: Wyoming big sagebrush, rabbitbrush, bluebunch wheatgrass, basin wildrye, Nevada bluegrass, Thurber needlegrass, and bottlebrush squirreltail

Occurrence in Idaho: southwestern part, mainly in Payette and Washington counties

Land Use: rangeland

The Prominent Characteristics of This Soil Are: clayey soil; wide cracks to the surface when dry; very slow permeability

The Ager soils have an interesting history of development from the original parent material of semiconsolidated tuff or siltstone. It weathered for many thousands of years. Clay that formed near the surface was slowly moved by water to lower depths in the soil. This process continued until the subsoil attained a moderately high content of clay. The soils at this stage of development were similar to **Soil 42 (Gem Series).** The specific kind of clay into which the soils developed has a high shrink-swell capacity. It swells when wetted and shrinks when dried.

As time went on, vertical cracks began to appear at the surface allowing granules of soil to fall into the cracks. This yearly wetting in winter months and drying in summer set up a churning action. Soil material pushed upwards from below as granules fell into the cracks. As the soils continued to weather, they became more and more clayey and the cracks became wider and deeper. This speeded the mixing process. These cracks are now 1 to 3 inches wide at the surface when the soil is dry.

The effects of this high shrink-swell and soil churning can easily be seen. Plants are easily destroyed. The power of the swelling when the soils are moistened is so great that foundations, sidewalks, and other structures are easily lifted out of place.

Another property of these soils which affects its use is the very slow permeability. Septic systems are doomed to failure. The soils are also extremely hard when dry.

Ager soils are problem soils for most any use.

Soil 54 (Boulder Lake Series)

A—0 to 25 inches; gray clay, prismatic structure, wide cracks, neutral. Upper inch has granular structure. Lower part has few slickensides.

AC—25 to 40 inches; light brownish gray silty clay loam, prismatic and subangular blocky structure, wide cracks, few slickensides, moderately alkaline.

Soil 54 (Boulder Lake Series) under natural vegetation of silver sagebrush.

Soil 54 (Boulder Lake Series)*

Family Classification: very fine, montmorillonitic, frigid Chromic Pelloxererts

Soil Depth: 60 inches or more

Drainage Class: somewhat poorly drained

Parent Material: lacustrine

Average Annual Precipitation: about 15 inches

Average Annual Air Temperature: about 43 degrees F.

Average Frost-Free Season: about 70 days

Elevation: about 5,700 feet

Topography: nearly level basins

Habitat Type: *Artemisia cana/Festuca idahoensis*

Common Native Vegetation: silver sagebrush, Idaho fescue, alkali bluegrass, hairgrass, sedge, rush, and cluster tarweed

Occurrence in Idaho: southwestern part, mainly in Owyhee county

Land Use: rangeland

The Prominent Characteristics of This Soil Are: clayey soil; wide cracks to the surface when dry; grayish colored soil horizons

This soil developed in silty lake-laid sediments. These sediments, coming from very old volcanic ash, settled into basins during a much wetter climate of the past.

After deposition of the parent material, many thousands of years passed with no erosion because of its topography. With a change to a drier climate, very little soil material has accumulated since the original deposition. This long period of stability allowed a special soil forming process to take place, similar to **Soil 52 (Magic Series)** and **Soil 53 (Ager Series).**

As weathering slowly took place, the percentage of clay gradually increased. Water passing downward through the soil carried along clay and deposited it below the surface soil. This increased the amount of clay in the lower depths even more. The kind of clay that formed has the property of expanding when water is available and shrinking when the soil dries. When enough clay accumulated near the surface vertical cracks began appearing in late summer. During this time granules of soil on the surface fell into the cracks. Once this process started, it continued to the present time, gobbling any small additions of alluvium which accumulated at the surface and slowly forcing up soil from below, creating a churning action.

This process will continue until about 7 inches of alluvium is added rather quickly to prevent the cracking to the surface. Another soil will then develop. This would be similar to what happened when **Soil 35 (Nez Perce Series)** formed.

The poor physical properties of this soil, in addition to being somewhat poorly drained, and having a short growing season, greatly limit its agricultural use. All building on this soil has to be carefully planned.

*The soil is outside the defined limits of the Boulder Lake Series because the A horizon is gray.

Bedrock (Granite)

Lichen-covered bedrock

Exposed bedrock in the Selkirk mountains of northern Idaho.

Bedrock (Granite)

Granite and granite-like rock are igneous rocks which solidified from molten material. They formed at possibly three different times about 60 to 100 million years ago in Idaho.

Magma from deep in the earth arose during those times toward the surface but remained at considerable depth where it cooled over long periods of time. This allowed for the development of large interlocking crystals which are visible in a piece of granite. This contrasts with fine grained basalt which is also igneous rock but which cooled very quickly at the surface of the earth.

After the magma cooled and became rock, the land area was then gradually uplifted which allowed overlying rock to erode, exposing the newly formed granite. The land continued to rise and mountains were formed. The granitic mountains are mostly in the higher elevations, about 6,000 to 11,000 feet, of the central part of the state and in the Priest Lake area of northern Idaho. Glaciers which were born much later in the ice age began their carving process which resulted in jagged mountain ridges and peaks. This resulted in thousands of acres of bare granite or related bedrock.

Weathering of the exposed bedrock is very slow in this coolest of climates in Idaho. The average annual air temperature is about 33 degrees F. Any material that is weathered is quickly eroded from the very steep and precipitous slopes. Trees which normally grow at these elevations cannot do so except for a few bonzai-like individuals which find a crack filled with some soil material. This granite bedrock then is not "soil" because it does not support land plants.

These granite bedrock areas yield a heavy runoff of snow water which is becoming more and more important for irrigation, recreation, and power production. It also is home to some species of wildlife. Backpackers frequent these rugged mountains in late summer to view the beautiful scenery.

Many other kinds of exposed bedrock occur throughout the state in all climates and elevations. An example is basalt, which is much younger than granite. The landcape photo of **Soil 41 (Tannahill Series)** shows basalt cliffs. This is a cutaway-view of a large number of basalt flows which came from fissures in the earth's crust and spread over large areas. Recent flows, occurring on the Snake River plains, can be seen in the background of the landscape of **Soil 22 (Moonville Series).**

Other general kinds of rocks are sedimentary rock and metamorphic rock. Sedimentary rock resulted from the consolidation of material deposited mainly by water. Metamorphic rock formed when igneous rock or sedimentary rock was altered by heat or pressure or by the introduction of magma or gases.

Whatever the kind of rock, it must undergo chemical and physical weathering to be a part of soil. This is done in the presence of heat and moisture over a long period of time. Rock slowly evolves to rocky soil material or residuum in which soils form. With further weathering, the nutrients which were locked up in the original rock can be taken up in solution by plant roots. With water and nutrients from this solution, along with carbon dioxide from the air and sunshine for photosynthesis, plant growth takes place.

The rocky soil material or residuum formed by weathering is subject to geologic erosion by wind and water. Tiny fragments of the original rock, which eventually make their way to a particular soil, may travel many miles. But they will not stay there forever. In time, all of todays soils will be completely eroded away. Sediment will fill seas of the future and the cycle will continue — rock to soil to rock...forever!

LITERATURE CITED

(1) United States Department of Agriculture. 1975. Soil Taxonomy: a basic system of soil classification for making and interpreting soil surveys. Soil Conservation Service, U.S. Dept. of Agric. Handbook 436, 754 pp. illus.

(2) United States Department of Agriculture. 1980. List of Scientific and Common Plant Names for Idaho. Soil Conservation Service, Boise, Idaho, 70 pp.

(3) United States Department of Agriculture. 1981. Soil Survey Manual. U.S. Dept. of Agric. Handbook 18, Chapter 4, Issue 1, 5/18, 105 pp.

APPENDIX

CONTENTS

Glossary 123
Soil classification (Taxonomy) 129
Depth to layer impeding root development 131
Natural soil drainage 132
Parent material 133
Average annual precipitation 135
Average annual air temperature 137
Average frost-free season 139
Elevation 141
Topography 143
Habitat type 144
List of plant names 146
Land use 147

GLOSSARY

The glossary includes words, phrases, and terms used by soil scientists when describing soils and explaining their genesis. Definitions and criteria were derived from several sources and have, in some cases, been tailored specifically for use in this atlas.

The table of soil classification is included for those with a knowledge of taxonomy or who have access to the publications listed on page 122. It is useful when comparing soils within the Idaho Soils Atlas as well as comparing soils in the atlas to the thousands of soils across the country.

Tables following the soil classification can be used to supplement the "Key to Prominent Soil Characteristics" (page 10). They are useful in comparing soils within the atlas having similar features, such as noting all the soils which formed in loess or those which have the same habitat type. The tables can also be used to locate soils in the atlas having a particular climate or topography.

The list of plant names gives the scientific name for the common name used in the atlas. Common plant names often vary from place to place.

A Soil Horizon (See Horizon, Soil)

Aggregate, Soil. Many fine particles held in a single mass or cluster. Natural soil aggregates, such as granules, blocks, or prisms, are called peds. Clods are man-caused aggregates produced mainly by tillage.

Alluvium. Material, such as sand, gravel, or silt transported and deposited on land by moving water.

Argillite. A rock derived either from siltstone or shale that has undergone a somewhat higher degree of induration than is present in those rocks.

Arid Climate. A climate that lacks sufficient moisture for crop production without irrigation.

Available Water Holding Capacity. The capacity of soils to hold water available for use by most plants. It is commonly defined as the difference between the amount of soil water at field moisture capacity and the amount at wilting point. The capacity, in inches, in a 60-inch profile or to a limiting layer is expressed as:

	Inches
VERY LOW	0 to 3
LOW	3 to 6
MODERATE	6 to 9
HIGH	9 to 12
VERY HIGH	12 or more

B Soil Horizon (See Horizon, Soil)

Bedrock. The solid rock that underlies the soil and other unconsolidated material or that is exposed at the surface.

Blocky Structure. (See Structure, Soil).

Bottomland. The normal flood plain of a stream, subject to frequent flooding.

Boulders. Rock fragments larger than 2 feet in diameter.

Bulk Density. The mass of dry soil per unit volume.

C Soil Horizon (See Horizon, Soil)

Calcareous Soil. Soil containing sufficient free calcium carbonate or calcium-magnesium carbonate to effervesce visibly when treated with cold 0.1 N hydrochloric acid.

Cirque. Semicircular, concave bowl-like area with steep faces primarily resulting from glacial ice and snow abrasion.

Clay. Mineral soil particles less than 0.002 millimeter in diameter.

Clay. (soil textural class). Soil material that is 40 percent or more clay, less than 45 percent sand, and less than 40 percent silt.

Clay Film. A thin coating of oriented clay on the surface of a soil aggregate, lining pores, or root channels.

Coarse Fragments. Mineral or rock particles 2 millimeters to 10 inches in diameter.

Coarse Textured Soil. Sand or loamy sand.

Cobblestone. A rounded or partly rounded fragment of rock 3 to 10 inches in diameter.

Cobbly. Material that is 15 to 35 percent, by volume, rock fragments, mostly cobblestone.

Cobbly, Very. Material that is 35 to 60 percent, by volume, rock fragments, mostly cobblestones.

Cobbly, Extremely. Material that is 60 percent or more, by volume, rock fragments, mostly cobblestones.

Colluvium. Soil material, rock fragments, or both, moved by creep, slide, or local wash and deposited on steep slopes.

Drainage Class. Refers to the frequency and duration of periods of saturation or partial saturation during soil formation, as opposed to altered drainage, which is commonly the result of artificial drainage or irrigation but may be caused by the sudden deepening of channels or the blocking of drainage outlets. Seven classes of natural soil drainage are:

Excessively Drained. Water is removed from the soil very rapidly. Excessively drained soils are commonly very coarse textured, rocky, or shallow. Some are steep. All are free of mottling related to wetness.

Somewhat Excessively Drained. Water is removed from the soil rapidly. Many somewhat excessively drained soils are sandy and rapidly permeable. Some are shallow. Some are so steep that much of the water they receive is lost as runoff. All are free of the mottling related to wetness.

Well Drained. Water is removed from the soil readily, but not rapidly. It is available to plants throughout most of the growing season, and wetness does not inhibit growth of roots for significant periods during most growing seasons. Well drained soils are commonly medium textured. They are mainly free of mottling.

Moderately Well Drained. Water is removed from the soil somewhat slowly during some periods. Moderately well drained soils are wet for only a short time during the growing season, but periodically for long enough that most crops are affected. They commonly have a slowly permeable layer within or directly below the solum, or periodically receive high rainfall or snowmelts, or both.

Somewhat Poorly Drained. Water is removed slowly enough that the soil is wet for significant periods during the growing season. Wetness markedly restricts the growth of most crops unless artificial drainage is provided. Somewhat poorly drained soils commonly have a slowly permeable layer, a high water table, additional water from seepage, or a combination of these.

Poorly Drained. Water is removed so slowly that the soil is saturated periodically during the growing season or remains wet for long periods. Free water is commonly at or near the surface for long enough during the growing season that most crops cannot be grown unless the soil is artificially drained. The soil is not continuously saturated in layers directly below plow depth. Poor drainage results from a high water table, a slowly permeable layer within the profile, seepage, or a combination of these.

Very Poorly Drained. Water is removed from the soil so slowly that free water remains at or on the surface during most of the growing season. Unless the soil is artificially drained, most crops cannot be grown. Very poorly drained soils are commonly level or depressed and are frequently ponded.

Duripan. A subsurface horizon that is cemented by silica. They commonly contain accessory cements, including calcium carbonate. They vary in appearance, but all have very firm moist consistence and are brittle even after prolonged wetting.

E Soil Horizon (See Horizon, Soil)

Effervescence. The formation of gaseous bubbles when a calcareous soil is treated with a diluted solution of hydrochloric acid.

Slightly effervescent - bubbles are readily observed.
Strongly effervescent - bubbles form a low foam.
Violently effervescent - thick foam forms quickly.

Eolian. Pertaining to material transported and deposited by the wind. Includes earthy materials ranging from dune sands to silty loess deposits.

Erosion. The wearing away of the land surface by water, wind, ice, or other geologic agents and by such processes as gravitational creep.

Erosion, Geologic. Erosion caused by geologic processes acting over long geologic periods and resulting in the wearing away of mountains and the building up of such landscape features as flood plains and coastal plains.

Erosion, Accelerated. Erosion much more rapid than geologic erosion, mainly as a result of the activities of man or other animals, or of a catastrophe in nature, for example, that exposes the surface.

Evapotranspiration. The combined loss of water by evaporation from the soil surface and by transpiration from plants.

Fine Textured Soil. Sandy clay, silty clay, and clay.

Fragipan. A loamy brittle subsurface horizon low in porosity and content of organic matter, and low or moderate in clay but high in silt or very fine sand. A fragipan appears cemented and restricts roots. When dry, it is hard or very hard and has a higher bulk density than the horizon or horizons above. When moist, it tends to rupture suddenly under pressure rather than to deform slowly.

Genesis, Soil. The mode of origin of the soil. Refers especially to the processes or soil-forming factors responsible for the formation of the solum from the parent material.

Glacial Moraine. A landform consisting of glacial till.

Glacial Outwash. Gravel, sand, and silt, commonly stratified, deposited by glacial melt water.

Glacial Till. Unsorted, nonstratified sediment, consisting of all sizes of material from clay to boulders, transported and deposited by glacial ice.

Glaciolacustrine. Material ranging from fine clay to sand derived from glaciers and deposited in glacial lakes mainly by glacial melt water. Many deposits are interbedded or laminated.

Granular Structure. (See Structure, Soil).

Gravel. Rounded or angular fragments of rock up to 3 inches in diameter.

Gravelly. Material that is 15 to 35 percent, by volume, rock fragments, mostly gravel.

Gravelly, Very. Material that is 35 to 60 percent, by volume, rock fragments, mostly gravel.

Gravelly, Extremely. Material that is 60 percent or more, by volume, rock fragments, mostly gravel.

Habitat Type. The collective area of land that supports or until recent time supported, and presumably is capable again of supporting, a particular climax plant association.

Horizon, Soil. A layer of soil, approximately parallel to the surface, having distinct characteristics produced by soil-forming processes. The major soil horizons are:

    A Horizon. The mineral horizon at or near the surface in which an accumulation of humified organic matter is mixed with mineral material. It is also a plowed surface horizon which had once been a B, C, or E soil horizon.

    B Horizon. The mineral horizon that formed below an A, E, or 0 horizon. It has (1) accumulated clay, iron, aluminum, humus, carbonates, gypsum, or silica, (2) evidence of removal of carbonates, (3) redder or browner colors from coatings of oxides than overlying or underlying horizons, (4) granular, blocky, or prismatic structure, or (5) any combinations of these.

    C Horizon. Horizon or layer, excluding hard bedrock, that is little effected by soil forming processes described in A, B, or E horizons.

    E Horizon. A mineral horizon which has lost a significant amount of clay, iron, or aluminum, leaving a concentration of sand and silt particles, mainly quartz.

    O Horizon. Layer dominated by organic material, such as undecomposed or partially decomposed leaves, needles, and twigs on the surface of mineral soils. Other O layers are in peat or muck which were deposited under water and that have decomposed to varying stages.

    R Layers. Hard bedrock.

Arabic numbers preceding the major horizons, starting with the number 2, indicate a significant change in particle-size distribution or mineralogy such as a difference in parent material. Where a soil has formed entirely in one kind of material, a prefix is omitted from the symbol; the whole profile is material 1.

Arabic numbers following the letters indicate a subdivision of a horizon.

Lower case letters are used as suffixes to designate specific kinds of master horizons or layers. The symbols are briefly described as follows:
    a - highly decomposed organic material
    b - buried genetic horizon
    c - concretions or hard nonconcretionary nodules
    e - organic material of intermediate decomposition
    g - strong gleying indicating either that iron has been reduced and removed during soil formation or that saturation with stagnant water has preserved a reduced state
    k - accumulation of carbonates
    m - cementation or induration
    n - accumulation of sodium
    p - plowing or other disturbance
    q - accumulation of silica
    r - weathered or soft bedrock
    s - accumulation of iron and aluminum oxides from above
    t - accumulation of clay
    w - development of color or structure
    x - fragipan character
    y - accumulation of gypsum
    z - accumulation of salts more soluble than gypsum

Igneous Rock. Rock formed by solidification of magma.

Infiltration. The downward entry of water into the immediate surface of soil or other material, as contrasted with percolation, which is movement of water through soil layers or material.

Irrigation. Application of water to soils to assist in production of crops.

Kaolinite. A type of clay mineral. It has 1:1 lattice structure (silica and aluminum).

Krotovinas. Irregular tubular streaks within one soil horizon of material transported from another horizon. They are caused by the filling of tunnels made by burrowing animals.

Lacustrine. Material deposited in lake water and exposed when the water level is lowered or the elevation of the land is raised.

Lamellae. Thin, wavy, generally horizontal, clayey bands mostly occurring in sandy soil material.

Lamination. A sedimentary layer less than 1 centimeter thick.

Leaching. The removal of soluble material from soil or other material by percolating water.

Lime. Calcium carbonate.

Limestone. A sedimentary rock consisting chiefly of calcium carbonate.

Loam. Soil material that is 7 to 27 percent clay particles, 28 to 50 percent silt particles, and less than 52 percent sand particles.

Loess. Fine grained material, dominantly of silt-sized particles, deposited by wind.

Magma. Naturally occurring molten rock material, generated within the earth and capable of intrusion and extrusion from which igneous rock are considered to have been derived by solidification.

Metamorphic Rock. Rock of any origin altered in mineralogical composition, chemical composition, or structure by heat, pressure, and movement at depth in the earth's crust. Nearly all such rocks are crystalline.

Mineral Soil. Soil that is mainly mineral material. Its bulk density is more than that of organic soil.

Moderately Coarse Textured Soil. Sandy loam and fine sandy loam.

Mottles. Spots or blotches of different color or shades of color interspersed with the dominant color. A common cause of mottling is impeded drainage, although there are other causes, such as soil development from an unevenly weathered rock.

O Soil Horizon (See Horizon, Soil)

Organic Matter. Plant and animal residue in the soil in various stages of decomposition.

Pan. A compact, dense layer in a soil that impedes the movement of water and the growth of roots. For example, duripan, fragipan, claypan, and plowpan.

Parent Material. The unconsolidated organic and mineral material in which soil forms.

Perched Water Table. The surface of a local zone of saturation held above the main body of groundwater by an impermeable layer or stratum, usually clay, and separated from the main body of groundwater by an unsaturated zone.

Percolation. The downward movement of water through the soil.

Permeability. The quality that enables the soil to transmit water or air, measured as the number of inches per hour that water moves downward through the saturated soil. Terms describing permeability are:

```
                                           Inches
VERY SLOW . . . . . . . . . . . . . . . . . . . . . . . less than 0.06
SLOW . . . . . . . . . . . . . . . . . . . . . . . . . . . . . .0.06 to 0.2
MODERATELY SLOW . . . . . . . . . . . . . . .0.2 to 0.6
MODERATE . . . . . . . . . . . . . . . . . . . . . . . . 0.6 to 2
MODERATELY RAPID . . . . . . . . . . . . . . . . . 2 to 6
RAPID . . . . . . . . . . . . . . . . . . . . . . . . . . . . . .6 to 20
VERY RAPID . . . . . . . . . . . . . . . . . . . . . . 20 or more
```

Platy Structure. (See Structure, Soil).

Plowpan. A compacted layer formed in the soil directly below the plowed layer.

Prismatic Structure. (See Structure, Soil).

Profile, Soil. A vertical section of the soil extending through all its horizons and into the parent material.

Quartzite. A metamorphic rock consisting essentially of quartz.

R Layers (See Horizon, Soil)

Rangeland. Land on which the potential natural vegetation is predominantly grasses, grasslike plants, forbs, or shrubs suitable for grazing or browsing. It includes natural grasslands, many wetlands, and areas that support forbs and shrubs.

Range Condition. The present composition of the plant community on a range site in relation to the potential natural plant community for that site. Range condition is expressed as excellent, good, fair, or poor, on the basis of how much the present plant community has departed from the potential.

Reaction, Soil. A measure of acidity or alkalinity of a soil, expressed in pH values. A soil that tests pH 7.0 is described as precisely neutral in reaction because it is neither acid nor alkaline. The degree of acidity or alkalinity is expressed as:

```
                                                    pH
EXTREMELY ACID . . . . . . . . . . . . . . . . . . . . .below 4.5
VERY STRONGLY ACID . . . . . . . . . . . . . . . .4.5 to 5.1
STRONGLY ACID . . . . . . . . . . . . . . . . . . . . . .5.1 to 5.6
MEDIUM ACID . . . . . . . . . . . . . . . . . . . . . . . .5.6 to 6.1
SLIGHTLY ACID . . . . . . . . . . . . . . . . . . . . . . .6.1 to 6.6
NEUTRAL . . . . . . . . . . . . . . . . . . . . . . . . . . . . .6.6 to 7.4
MILDLY ALKALINE . . . . . . . . . . . . . . . . . . . . .7.4 to 7.9
MODERATELY ALKALINE . . . . . . . . . . . . . .7.9 to 8.5
STRONGLY ALKALINE . . . . . . . . . . . . . . . . .8.5 to 9.1
VERY STRONGLY ALKALINE . . . . . . . . . .9.1 and higher
```

Residuum. Unconsolidated, weathered, or partly weathered mineral material that accumulated as consolidated rock disintegrated in place.

Rippable. Bedrock or duripan that can be excavated using a single-tooth ripping attachment mounted on a tractor with a 200-300 draw bar horsepower rating.

Rock Fragments. Rock or mineral fragments having a diameter of 2 millimeters or more; includes gravel, cobblestones, stones, and boulders.

Sand. Individual rock or mineral fragments from 0.05 millimeter to 2.0 millimeters in diameter. Most sand grains are quartz.

Sand (soil textural class). A soil that is 85 percent or more sand and not more than 10 percent clay.

Sandstone. Sedimentary rock containing dominantly sand-size particles.

Sapric Materials. Highly decomposed organic materials, largely plant remains.

Scree. A heap of rock waste at the base of a cliff or a sheet of coarse debris mantling a mountain slope.

Sedimentary Rock. Rock made up of particles deposited from suspension in water. The chief kinds of sedimentary rock are sandstone, formed from sand; siltstone, formed from silt; shale, formed from clay; and limestone, formed from soft masses of calcium carbonate. There are many intermediate types. Some wind-deposited sand is consolidated into sandstone.

Series, Soil. A group of soils that have about the same profile, except for differences in texture of the surface layer or of the underlying material. All the soils of a series have horizons that are similar in composition, thickness, and arrangement.

Shale. A sedimentary rock formed by induration of a clay or silty clay deposit and having the tendency to split into thin layers.

Shrink-Swell. The shrinking of soil when dry and the swelling when wet. Shrinking and swelling can damage roads, dams, building foundations, and other structures. It can also damage plant roots.

Silica. A combination of silicon and oxygen. The mineral form is called quartz.

Silt. Individual mineral particles that range in diameter from the upper limit of clay (0.002 millimeter) to the lower limit of very fine sand (0.05 millimeter).

Silt (soil textural class). Soil that is 80 percent or more silt and less than 12 percent clay.

Siltstone. A sedimentary rock composed predominantly of silt-sized particles.

Slick Spot. A small area of soil having a puddled, crusted, or smooth surface caused by an excess of sodium. The soil is generally silty or clayey, is slippery when wet, and is low in productivity.

Slickensides. Polished and grooved surfaces that are produced by one mass sliding past another. They are very common in swelling clays in which there are marked changes in moisture content.

Soil. A natural, three-dimensional body at the earth's surface. It is capable of supporting land plants and has properties resulting from the integrated effect of climate and living matter acting on earthy parent material, as conditioned by relief over periods of time.

Soil Depth. The depth of a soil to a layer which essentially inhibits root growth — bedrock, duripan, fragipan. Depth groupings are:

	Inches
VERY DEEP	60 or more
DEEP	40 to 60
MODERATELY DEEP	20 to 40
SHALLOW	10 to 20
VERY SHALLOW	less than 10

Soil Particles. Mineral particles less than 2 mm in diameter and ranging between specified size limits. The names and sizes of separates are:

	Millimeters
VERY COARSE SAND	2.0 to 1.0
COARSE SAND	1.0 to 0.5
MEDIUM SAND	0.5 to 0.25
FINE SAND	0.25 to 0.10
VERY FINE SAND	0.10 to 0.05
SILT	0.05 to 0.002
CLAY	less than 0.002

Soil Slope. Expressed in terms of percentage — the difference in elevation in feet for each 100 feet horizontal distance. Normally each slope class has variable limits but those chosen for this atlas are:

	Slope (percent)
NEARLY LEVEL	0 to 2
GENTLY SLOPING	2 to 6
MODERATELY SLOPING (OR ROLLING)	6 to 12
MODERATELY SLOPING (OR HILLY)	12 to 25
STEEP	25 to 50
VERY STEEP	50 and more

Solum. The upper part of a soil profile, above the C horizon, in which the processes of soil formation are active. The solum in soil consists of the A and B horizons. Generally, the characteristics of the material in these horizons are unlike those of the underlying material. The living roots and plant and animal activities are largely confined to the solum.

Stones. Rock fragments 10 to 24 inches in diameter.

Stony, Very. Material that is 35 to 60 percent, by volume, rock fragments, mostly stones.

Structure, Soil. The arrangement of primary soil particles into compound particles or aggregates. The principal forms of soil structure are: platy (laminated), prismatic (vertical axis of aggregates longer than horizontal), blocky (angular or subangular), and granular. Structureless soils are either single grained (each grain by itself, as in dune sand) or massive (the particles adhering without any regular cleavage, as in many duripans).

Subsoil. Technically, the B horizon; roughly, the part of the solum below plow depth.

Substratum. The part of the soil below the solum.

Subsurface Layer. Technically, the E horizon. Generally refers to a leached horizon lighter in color and lower in content of organic matter than the overlying surface layer.

Summer Fallow. The tillage of uncropped land during the summer to control weeds and allow storage of moisture in the soil for the growth of a later crop. A practice common in semiarid regions, where annual precipitation is not enough to produce a crop every year. Summer fallow is frequently practiced before planting winter grain.

Surface Layer. The original or present A horizon, including the transitional AB horizon, varying widely among different soils. Usually it is the dark colored upper soil that ranges from less than 1 inch to 2 feet or more in thickness on different soils.

Surface Soil. The soil ordinarily moved in tillage, or its equivalent in uncultivated soil, ranging in depth from 4 to 10 inches. Frequently designated as the "plow layer" or the "Ap horizon".

Terrace (Geologic). An old alluvial plain, ordinarily flat or undulating, generally bordering a river or a lake.

Texture, Soil. The relative proportions of sand, silt, and clay particles in a mass of soil. The basic textural classes, in order of increasing proportion of fine particles, are: sand, loamy sand, sandy loam, loam, silt, silt loam, sandy clay loam, clay loam, silty clay loam, sandy clay, silty clay, and clay. The sand, loamy sand, and sandy loam classes may be further divided by specifying "coarse", "fine", or "very fine".

Tilth, Soil. The condition of the soil, especially the soil structure, as related to the growth of plants. Good tilth refers to the friable state and is associated with high noncapillary porosity and stable structure. A soil in poor tilth is nonfriable, hard, nonaggregated, and difficult to till.

Tuff. A rock formed of compact volcanic fragments, generally smaller than 4 millimeters in diameter.

Uncoated Silt Grains. Silt particles which have had coatings of organic matter, oxides, or other material removed by leaching.

Upland. Land at a higher elevation, in general, than the alluvial plain or stream terrace; land above the low lands along streams.

Variant, Soil. A soil having properties sufficiently different from those of other known soils to justify a new series name, but occurring in such a limited geographic area that creation of a new series is not justified.

SOIL CLASSIFICATION (TAXONOMY)(1)

SOIL	SOIL ORDER / SOIL FAMILY
	ALFISOLS
Soil 1	Fine-loamy, mixed Typic Cryoboralfs
Soil 2 (Santa Series)	Coarse-silty, mixed, frigid Ochreptic Fragixeralfs
Soil 3 (Porthill Series)	Fine, mixed, frigid Typic Haploxeralfs
	ARIDISOLS
Soil 4 (Chilcott Series)	Fine, montmorillonitic, mesic Abruptic Xerollic Durargids
Soil 5 (Colthorp Series)	Loamy, mixed, mesic, shallow Xerollic Durargids
Soil 6 (Sebree Series)	Fine-silty, mixed, mesic Xerollic Nadurargids
Soil 7 (Gooding Series)	Fine, montmorillonitic, mesic Xerollic Paleargids
Soil 8 (Portneuf Series)	Coarse-silty, mixed, mesic Durixerollic Calciorthids
Soil 9 (Trevino Series)	Loamy, mixed, mesic Lithic Xerollic Camborthids
Soil 10 (Owhyee Series)	Coarse-silty, mixed, mesic Xerollic Camborthids
Soil 11 (Minidoka Series)	Coarse-silty, mixed, mesic Xerollic Durorthids
Soil 12	Loamy, mixed, frigid Xerollic Durorthids
	ENTISOLS
Soil 13 (Garbutt Series)	Coarse-silty, mixed (calcareous), mesic Typic Torriorthents
Soil 14 (Flybow Series)	Loamy-skeletal, mixed, monacid, mesic Lithic Xerorthents
Soil 15 (Pyle Series)	Mixed Alfic Cryopsamments
Soil 16 (Quincy Series)	Mixed, mesic Xeric Torripsamments
Soil 17 (Shellrock Series)	Mixed, frigid Typic Xeropsamments
	HISTISOLS
Soil 18 (Pywell Series)	Euic Typic Borosaprists
	INCEPTISOLS
Soil 19 (Vay Series)	Medial over loamy-skeletal, mixed Entic Cryandepts
Soil 20 (Bluehill Series)	Ashy, mesic Typic Vitrandepts
Soil 21	Medial over loamy, mixed, frigid Typic Vitrandepts
Soil 22 (Moonville Series)	Cindery, frigid Mollic Vitrandepts
Soil 23 (Roseberry Series)	Sandy, mixed Humic Cryaquepts
Soil 24	Loamy-skeletal, mixed Andic Cryochrepts
Soil 25	Loamy-skeletal, mixed Dystric Cryochrepts
Soil 26	Fine-loamy, mixed Dystric Cryochrepts
Soil 27	Coarse-loamy, mixed, frigid Andic Dystrochrepts
Soil 28 (Moonville Variant)	Coarse-loamy, mixed, frigid Andic Xerochrepts
Soil 29 (Bonner Series)	Coarse-loamy over sandy or sandy skeletal, mixed, frigid Andic Xerochrepts

Soil 30 (Oxford Series)	Fine, montmorillonitic, frigid Vertic Xerochrepts
Soil 31 (McCall Series)	Loamy-skeletal, mixed Typic Cryumbrepts
Soil 32	Coarse-loamy, mixed Andic Cryumbrepts

MOLLISOLS

Soil 33 (Southwick Series)	Fine-silty, mixed, mesic Argiaquic Xeric Argialbolls
Soil 34 (Houk Series)	Fine, montmorillonitic, frigid Argiaquic Xeric Argialbolls
Soil 35 (Nez Perce Series)	Fine, montmorillonitic, mesic Xeric Argialbolls
Soil 36 (Driggs Variant)	Fine-loamy, mixed Argic Cryoborolls
Soil 37	Fine-loamy, mixed Argic Pachic Cryoborolls
Soil 38 (Greys Series)	Fine-silty, mixed Boralfic Cryoborolls
Soil 39	Very fine, mixed Duric Cryoborolls
Soil 40 (Pavohroo Series)	Fine-loamy, mixed Pachic Cryoborolls
Soil 41 (Tannahill Series)	Loamy-skeletal, mixed mesic Calcic Argixerolls
Soil 42 (Gem Series)	Fine, montmorillonitic, mesic Calcic Argixerolls
Soil 43 (Gwin Series)	Loamy-skeletal, mixed, mesic Lithic Argixerolls
Soil 44 (Klickson Series)	Loamy-skeletal, mixed, frigid Ultic Argixerolls
Soil 45 (Little Wood Series)	Loamy-skeletal, mixed, frigid Ultic Argixerolls
Soil 46 (Rexburg Series)	Coarse-silty, mixed, frigid Calcic Haploxerolls
Soil 47 (Westlake Series)	Fine-silty, mixed, frigid Cumulic Ultic Haploxerolls
Soil 48 (Hymas Series)	Loamy-skeletal, carbonantic, frigid Lithic Haploxerolls
Soil 49 (Ola Series)	Coarse-loamy, mixed, frigid Pachic Haploxerolls
Soil 50 (Palouse Series)	Fine-silty, mixed, mesic Pachic Ultic Haploxerolls

SPODOSOLS

Soil 51	Coarse-loamy, mixed Entic Cryorthods

VERTISOLS

Soil 52 (Magic Series)	Fine, montmorillonitic, frigid Entic Chromoxererts
Soil 53 (Ager Series)	Very fine, montmorillonitic, mesic Entic Chromoxererts
Soil 54 (Boulder Lake Series)	Very fine, montmorillonitic, frigid Chromic Pelloxererts

DEPTH TO LAYER IMPEDING ROOT DEVELOPMENT

SOIL DEPTH (inches)	SOIL
60 or more	Soil 1
	Soil 3 (Porthill Series)
	Soil 10 (Owyhee Series)
	Soil 13 (Garbutt Series)
	Soil 16 (Quincy Series)
	Soil 18 (Pywell Series)
	Soil 21
	Soil 22 (Moonville Series)
	Soil 23 (Roseberry Series)
	Soil 24
	Soil 25
	Soil 26
	Soil 28 (Moonville Variant)
	Soil 29 (Bonner Series)
	Soil 30 (Oxford Series)
	Soil 31 (McCall Series)
	Soil 32
	Soil 33 (Southwick Series)
	Soil 34 (Houk Series)
	Soil 35 (Nez Perce Series)
	Soil 36 (Driggs Variant)
	Soil 37
	Soil 38 (Greys Series)
	Soil 40 (Pavohroo Series)
	Soil 44 (Klickson Series)
	Soil 45 (Little Wood Series)
	Soil 46 (Rexburg Series)
	Soil 47 (Westlake Series)
	Soil 50 (Palouse Series)
	Soil 51
	Soil 53 (Ager Series)
	Soil 54 (Boulder Lake Series)
40 to 60 or more	Soil 7 (Gooding Series)
	Soil 19 (Vay Series)
	Soil 41 (Tannahill Series)
40 to 60	Soil 17 (Shellrock Series)
	Soil 27
20 to 40	Soil 2 (Santa Series)
	Soil 4 (Chilcott Series)
	Soil 6 (Sebree Series)
	Soil 11 (Minidoka Series)
	Soil 15 (Pyle Series)
	Soil 20 (Bluehill Series)
	Soil 39
	Soil 42 (Gem Series)
	Soil 49 (Ola Series)
	Soil 52 (Magic Series)
10 to 20	Soil 5 (Colthorp Series)
	Soil 9 (Trevino Series)
	Soil 12
	Soil 43 (Gwin Series)
	Soil 48 (Hymas Series)
2 to 10	Soil 14 (Flybow Series)

NATURAL SOIL DRAINAGE

DRAINAGE CLASS	SOIL
Very poorly drained	**Soil 18 (Pywell Series)**
Poorly drained	**Soil 23 (Roseberry Series)**
Somewhat poorly drained	**Soil 34 (Houk Series)** **Soil 47 (Westlake Series)** **Soil 54 (Boulder Lake Series)**
Moderately well drained	**Soil 2 (Santa Series)** **Soil 3 (Porthill Series)** **Soil 33 (Southwick Series)** **Soil 35 (Nez Perce Series)** **Soil 51**
Well drained	**Soil 1** **Soil 4 (Chilcott Series)** **Soil 5 (Colthorp Series)** **Soil 6 (Sebree Series)** **Soil 7 (Gooding Series)** **Soil 8 (Portneuf Series)** **Soil 9 (Trevino Series)** **Soil 10 (Owyhee Series)** **Soil 11 (Minidoka Series)** **Soil 12** **Soil 13 (Garbutt Series)** **Soil 14 (Flybow Series)** **Soil 19 (Vay Series)** **Soil 21** **Soil 24** **Soil 25** **Soil 26** **Soil 27** **Soil 28 (Moonville Variant)** **Soil 29 (Bonner Series)** **Soil 30 (Oxford Series)** **Soil 32** **Soil 36 (Driggs Variant)** **Soil 37** **Soil 38 (Greys Series)** **Soil 39** **Soil 40 (Pavohroo Series)** **Soil 41 (Tannahill Series)** **Soil 42 (Gem Series)** **Soil 43 (Gwin Series)** **Soil 44 (Klickson Series)** **Soil 45 (Little Wood Series)** **Soil 46 (Rexburg Series)** **Soil 48 (Hymas Series)** **Soil 49 (Ola Series)** **Soil 50 (Palouse Series)** **Soil 52 (Magic Series)** **Soil 53 (Ager Series)**
Somewhat excessively drained	**Soil 15 (Pyle Series)** **Soil 17 (Shellrock Series)** **Soil 20 (Bluehill Series)** **Soil 22 (Moonville Series)** **Soil 31 (McCall Series)**
Excessively drained	**Soil 16 (Quincy Series)**

PARENT MATERIAL

TYPE	SOIL
Alluvium	**Soil 12** **Soil 13 (Garbutt Series)** **Soil 34 (Houk Series)** **Soil 36 (Driggs Variant)** **Soil 45 (Little Wood Series)** **Soil 47 (Westlake Series)**
Alluvium over loess over basalt residuum	**Soil 7 (Gooding Series)**
Alluvium over rhyolitic welded tuff	**Soil 39**
Basalt residuum	**Soil 14 (Flybow Series)** **Soil 26** **Soil 42 (Gem Series)** **Soil 52 (Magic Series)**
Eolian sand	**Soil 16 (Quincy Series)**
Glacial outwash	**Soil 23 (Roseberry Series)** **Soil 37**
Glacial till	**Soil 31 (McCall Series)**
Glaciolacustrine	**Soil 3 (Porthill Series)**
Granite, gneiss, granodiorite, quartz monzonite, or quartz diorite residuum	**Soil 15 (Pyle Series)** **Soil 17 (Shellrock Series)** **Soil 49 (Ola Series)**
Lacustrine	**Soil 10 (Owyhee Series)** **Soil 30 (Oxford Series)** **Soil 54 (Boulder Lake Series)**
Limestone residuum	**Soil 48 (Hymas Series)**
Loess	**Soil 2 (Santa Series)** **Soil 8 (Portneuf Series)** **Soil 33 (Southwick Series)** **Soil 35 (Nez Perce Series)** **Soil 38 (Greys Series)** **Soil 46 (Rexburg Series)** **Soil 50 (Palouse Series)**
Loess and alluvium	**Soil 4 (Chilcott Series)** **Soil 6 (Sebree Series)**
Loess mixed with basalt	**Soil 9 (Trevino Series)**
Loess mixed with basalt colluvium	**Soil 41 (Tannahill Series)** **Soil 43 (Gwin Series)** **Soil 44 (Klickson Series)**
Loess or silty alluvium	**Soil 11 (Minidoka Series)**
Loess over limestone residuum	**Soil 1** **Soil 40 (Pavohroo Series)**
Loess over basalt bedrock	**Soil 5 (Colthorp Series)**
Organic material	**Soil 18 (Pywell Series)**
Quartzite colluvium	**Soil 25**
Tuff or siltstone residuum	**Soil 53 (Ager Series)**
Volcanic ash	**Soil 20 (Bluehill Series)**

Volcanic ash and cinders	**Soil 22 (Moonville Series)** **Soil 28 (Moonville Variant)**
Volcanic ash mixed with glacial till	**Soil 24**
Volcanic ash mixed with material weathered from granite or gneiss	**Soil 19 (Vay Series)** **Soil 27** **Soil 32**
Volcanic ash over glacial outwash	**Soil 29 (Bonner Series)**
Volcanic ash over glacial till	**Soil 51**
Volcanic ash over schist residuum	**Soil 21**

AVERAGE ANNUAL PRECIPITATION

INCHES	SOIL
8	Soil 13 (Garbutt Series) Soil 16 (Quincy Series)
9	Soil 4 (Chilcott Series) Soil 7 (Gooding Series) Soil 8 (Portneuf Series) Soil 9 (Trevino Series) Soil 10 (Owyhee Series) Soil 11 (Minidoka Series)
10	Soil 6 (Sebree Series) Soil 12
11	Soil 5 (Colthorp Series)
12	Soil 53 (Ager Series)
13	Soil 48 (Hymas Series)
14	Soil 20 (Bluehill Series) Soil 22 (Moonville Series) Soil 28 (Moonville Variant) Soil 34 (Houk Series) Soil 41 (Tannahill Series) Soil 42 (Gem Series) Soil 45 (Little Wood Series) Soil 46 (Rexburg Series) Soil 52 (Magic Series)
15	Soil 36 (Driggs Variant) Soil 54 (Boulder Lake Series)
16	Soil 39
17	Soil 30 (Oxford Series) Soil 37
18	Soil 49 (Ola Series)
19	Soil 14 (Flybow Series) Soil 38 (Greys Series)
20	Soil 43 (Gwin Series)
21	Soil 3 (Porthill Series) Soil 50 (Palouse Series)
22	Soil 35 (Nez Perce Series) Soil 47 (Westlake Series)
23	Soil 23 (Roseberry Series) Soil 33 (Southwick Series)
24	Soil 25 Soil 40 (Pavohroo Series)
25	Soil 1 Soil 18 (Pywell Series) Soil 31 (McCall Series)
26	Soil 44 (Klickson Series)
27	Soil 17 (Shellrock Series)
28	Soil 2 (Santa Series)
30	Soil 15 (Pyle Series) Soil 26 Soil 29 (Bonner Series)

40	**Soil 19 (Vay Series)**
	Soil 21
	Soil 27
50	**Soil 24**
	Soil 32
60	**Soil 51**

AVERAGE ANNUAL AIR TEMPERATURE

DEGREE F.	SOIL
35	Soil 32
36	Soil 25
	Soil 51
37	Soil 37
38	Soil 1
	Soil 24
39	Soil 26
	Soil 31 (McCall Series)
	Soil 38 (Greys Series)
40	Soil 15 (Pyle Series)
	Soil 19 (Vay Series)
	Soil 23 (Roseberry Series)
	Soil 36 (Driggs Variant)
	Soil 39
	Soil 40 (Pavohroo Series)
41	Soil 12
	Soil 17 (Shellrock Series)
	Soil 34 (Houk Series)
	Soil 52 (Magic Series)
42	Soil 3 (Porthill Series)
	Soil 27
	Soil 28 (Moonville Variant)
	Soil 44 (Klickson Series)
43	Soil 2 (Santa Series)
	Soil 29 (Bonner Series)
	Soil 45 (Little Wood Series)
	Soil 46 (Rexburg Series)
	Soil 48 (Hymas Series)
	Soil 54 (Boulder Lake Series)
44	Soil 21
	Soil 22 (Moonville Series)
	Soil 30 (Oxford Series)
	Soil 47 (Westlake Series)
	Soil 49 (Ola Series)
45	Soil 18 (Pywell Series)
46	Soil 20 (Bluehill Series)
	Soil 33 (Southwick Series)
	Soil 35 (Nez Perce Series)
47	Soil 6 (Sebree Series)
	Soil 9 (Trevino Series)
	Soil 42 (Gem Series)
48	Soil 8 (Portneuf Series)
	Soil 14 (Flybow Series)
	Soil 43 (Gwin Series)
	Soil 50 (Palouse Series)
	Soil 53 (Ager Series)
49	Soil 7 (Gooding Series)
50	Soil 11 (Minidoka Series)
51	Soil 4 (Chilcott Series)
	Soil 5 (Colthorp Series)
	Soil 13 (Garbutt Series)

Soil 10 (Owyhee Series)
Soil 16 (Quincy Series)
Soil 41 (Tannahill Series)

AVERAGE FROST-FREE SEASON

DAYS	SOIL
50 or less	Soil 1
	Soil 15 (Pyle Series)
	Soil 24
	Soil 25
	Soil 32
	Soil 37
	Soil 39
	Soil 40 (Pavohroo Series)
	Soil 51
60	Soil 26
70	Soil 23 (Roseberry Series)
	Soil 31 (McCall Series)
	Soil 38 (Greys Series)
	Soil 54 (Boulder Lake Series)
75	Soil 19 (Vay Series)
	Soil 21
	Soil 22 (Moonville Series)
	Soil 27
80	Soil 12
	Soil 17 (Shellrock Series)
	Soil 28 (Moonville Variant)
	Soil 36 (Driggs Variant)
	Soil 44 (Klickson Series)
	Soil 48 (Hymas Series)
85	Soil 34 (Houk Series)
	Soil 45 (Little Wood Series)
	Soil 49 (Ola Series)
	Soil 52 (Magic Series)
90	Soil 46 (Rexburg Series)
95	Soil 18 (Pywell Series)
	Soil 30 (Oxford Series)
100	Soil 2 (Santa Series)
	Soil 47 (Westlake Series)
105	Soil 20 (Bluehill Series)
110	Soil 29 (Bonner Series)
115	Soil 33 (Southwick Series)
120	Soil 35 (Nez Perce Series)
125	Soil 3 (Porthill Series)
	Soil 7 (Gooding Series)
	Soil 42 (Gem Series)
130	Soil 8 (Portneuf Series)
	Soil 53 (Ager Series)
135	Soil 5 (Colthorp Series)
	Soil 9 (Trevino Series)
	Soil 11 (Minidoka Series)
	Soil 13 (Garbutt Series)
140	Soil 6 (Sebree Series)
	Soil 43 (Gwin Series)
	Soil 50 (Palouse Series)
150	Soil 4 (Chilcott Series)
	Soil 10 (Owyhee Series)

Soil 14 (Flybow Series)
Soil 16 (Quincy Series)
Soil 41 (Tannahill Series)

ELEVATION

AVERAGE ELEVATION (feet)	SOIL
1,800	Soil 41 (Tannahill Series)
2,200	Soil 18 (Pywell Series)
2,300	Soil 3 (Porthill Series)
2,500	Soil 10 (Owyhee Series) Soil 16 (Quincy Series) Soil 29 (Bonner Series)
2,700	Soil 14 (Flybow Series) Soil 50 (Palouse Series)
2,800	Soil 2 (Santa Series) Soil 53 (Ager Series)
2,900	Soil 33 (Southwick Series) Soil 43 (Gwin Series)
3,000	Soil 4 (Chilcott Series)
3,400	Soil 35 (Nez Perce Series)
3,500	Soil 5 (Colthorp Series) Soil 6 (Sebree Series) Soil 7 (Gooding Series) Soil 9 (Trevino Series) Soil 11 (Minidoka Series) Soil 21 Soil 44 (Klickson Series) Soil 47 (Westlake Series)
3,600	Soil 8 (Portneuf Series)
3,700	Soil 13 (Garbutt Series) Soil 27
3,800	Soil 42 (Gem Series)
4,000	Soil 49 (Ola Series)
4,100	Soil 19 (Vay Series)
4,400	Soil 23 (Roseberry Series)
4,900	Soil 30 (Oxford Series)
5,000	Soil 26
5,100	Soil 31 (McCall Series) Soil 34 (Houk Series) Soil 45 (Little Wood Series)
5,200	Soil 52 (Magic Series)
5,300	Soil 20 (Bluehill Series)
5,400	Soil 22 (Moonville Series) Soil 46 (Rexburg Series)
5,500	Soil 17 (Shellrock Series) Soil 28 (Moonville Variant)
5,700	Soil 24 Soil 54 (Boulder Lake Series)
5,800	Soil 48 (Hymas Series) Soil 51

6,000	**Soil 15 (Pyle Series)**
6,200	**Soil 32**
6,300	**Soil 36 (Driggs Variant)**
	Soil 38 (Greys Series)
	Soil 39
6,500	**Soil 12**
	Soil 40 (Pavohroo Series)
7,000	**Soil 25**
7,300	**Soil 1**
7,500	**Soil 37**

TOPOGRAPHY

TYPE	SOIL
Nearly level bottom lands, terraces, or basins	**Soil 18 (Pywell Series)** **Soil 23 (Roseberry Series)** **Soil 34 (Houk Series)** **Soil 47 (Westlake Series)** **Soil 54 (Boulder Lake Series)**
Nearly level and gently sloping fans, terraces, or plains	**Soil 3 (Porthill Series)** **Soil 10 (Owyhee Series)** **Soil 11 (Minidoka Series)** **Soil 13 (Garbutt Series)** **Soil 28 (Moonville Variant)** **Soil 29 (Bonner Series)** **Soil 45 (Little Wood Series)** **Soil 52 (Magic Series)**
Nearly level to moderately sloping or rolling plains	**Soil 7 (Gooding Series)** **Soil 22 (Moonville Series)** **Soil 39** **Soil 6 (Sebree Series)**
Gently sloping plains	**Soil 4 (Chilcott Series)** **Soil 5 (Colthorp Series)**
Gently and moderately sloping terraces or plains	**Soil 8 (Portneuf Series)** **Soil 9 (Trevino Series)** **Soil 12** **Soil 36 (Driggs Variant)** **Soil 37**
Gently sloping to moderately steep cirque basins or dissected lakebed terraces	**Soil 30 (Oxford Series)** **Soil 51**
Gently sloping to hilly uplands	**Soil 2 (Santa Series)** **Soil 33 (Southwick Series)** **Soil 35 (Nez Perce Series)** **Soil 38 (Greys Series)** **Soil 50 (Palouse Series)**
Gently sloping to steep hummocks or dunes	**Soil 16 (Quincy Series)**
Gently sloping to steep uplands	**Soil 26** **Soil 31 (McCall Series)** **Soil 42 (Gem Series)** **Soil 46 (Rexburg Series)** **Soil 53 (Ager Series)**
Moderately sloping to steep ridges, foothills, canyons, or mountains	**Soil 1** **Soil 14 (Flybow Series)** **Soil 17 (Shellrock Series)** **Soil 43 (Gwin Series)** **Soil 48 (Hymas Series)**
Moderately steep to very steep hills, foothills, canyons, or mountains	**Soil 20 (Bluehill Series)** **Soil 40 (Pavohroo Series)** **Soil 41 (Tannahill Series)** **Soil 44 (Klickson Series)** **Soil 49 (Ola Series)**
Steep and very steep foothills and/or mountains	**Soil 15 (Pyle Series)** **Soil 19 (Vay Series)** **Soil 21** **Soil 24** **Soil 25** **Soil 27** **Soil 32**

HABITAT TYPE

TYPE	SOIL
Abies grandis/Pachystima myrsinites	**Soil 2 (Santa Series)** **Soil 29 (Bonner Series)**
Abies grandis/Spiraea betulifolia	**Soil 26**
Abies grandis/Vaccinium membranaceum	**Soil 31 (McCall Series)**
Abies lasiocarpa/Menziesia ferruginea	**Soil 51**
Abies lasiocarpa/Pachystima myrsinites	**Soil 1**
Abies lasiocarpa/Xerophyllum tenax	**Soil 24** **Soil 25** **Soil 32**
Agropyron spicatum/Opuntia polyacantha	**Soil 14 (Flybow Series)**
Agropyron spicatum/Poa sandbergii	**Soil 41 (Tannahill Series)**
Artemisia arbuscula/Festuca idahoensis	**Soil 39**
Artemisia arbuscula/Agropyron spicatum	**Soil 12**
Artemisia cana/Festuca idahoensis	**Soil 34 (Houk Series)** **Soil 54 (Boulder Lake Series)**
Artemisia longiloba/Festuca idahoensis	**Soil 52 (Magic Series)**
Artemisia nova/Agropyron spicatum	**Soil 48 (Hymas Series)**
Artemisia tridentata tridentata/Oryzopsis hymenoides	**Soil 16 (Quincy Series)**
Artemisia tridentata vaseyana/Agropyron spicatum	**Soil 30 (Oxford Series)**
Artemisia tridentata vaseyana/Festuca idahoensis idahoensis	**Soil 22 (Moonville Series)** **Soil 28 (Moonville Variant)** **Soil 36 (Driggs Variant)** **Soil 45 (Little Wood Series)** **Soil 49 (Ola Series)**
Artemisia tridentata vaseyana spiciformis/Festuca	**Soil 37**
Artemisia tridentata xericensis/Agropyron spicatum	**Soil 42 (Gem Series)** **Soil 4 (Chilcott Series)** **Soil 5 (Colthorp Series)** **Soil 7 (Gooding Series)** **Soil 9 (Trevino Series)** **Soil 10 (Owyhee Series)** **Soil 53 (Ager Series)**
Artemisia tridentata wyomingensis/Oryzopsis hymenoides	**Soil 20 (Bluehill Series)**
Artemisia tridentata wyomingensis/Stipa thurberiana	**Soil 8 (Portneuf Series)** **Soil 11 (Minidoka Series)**
Artemisia tripartita/Agropyron spicatum	**Soil 46 (Rexburg Series)**
Eurotia lanata/Oryzopsis hymenoides	**Soil 13 (Garbutt Series)**
Festuca idahoensis/Agropyron spicatum	**Soil 43 (Gwin Series)**
Festuca idahoensis/Rosa nutkana	**Soil 35 (Nez Perce Series)**
Festuca idahoensis/Symphoricarpos albus	**Soil 50 (Palouse Series)**
Pinus ponderosa/Purshia tridentata, Festuca idahoensis phase	**Soil 17 (Shellrock Series)**

Pinus ponderosa/Symphoricarpos albus	**Soil 33 (Southwick Series)**
Populus tremuloides/Calamagrostis rubescens	**Soil 38 (Greys Series)**
Psuedotsuga menziesii/Calamagrostis rubescens	**Soil 40 (Pavohroo Series)**
Pseudotsuga menziesii/Physocarpus malvaceus	**Soil 15 (Pyle Series)** **Soil 44 (Klickson Series)**
Thuja plicata/Pachystima myrsinites	**Soil 3 (Porthill Series)** **Soil 19 (Vay Series)** **Soil 21** **Soil 27**
Barren areas within *Artemisia tridentata wyomingensis/Agropyron spicatum*	**Soil 6 (Sebree Series)**
Unclassified	**Soil 18 (Pywell Series)** **Soil 23 (Roseberry Series)** **Soil 47 (Westlake Series)**

LIST OF PLANT NAMES (2)

COMMON NAME	SCIENTIFIC NAME
Alkali bluegrass	Poa juncifolia
Alkali sagebrush	Artemisia longiloba
American trailplant	Adenocaulon bicolor
Antelope bitterbrush	Purshia tridentata
Arrowleaf balsamroot	Balsamorhiza sagittata
Baltic rush	Juncus balticus
Basin big sagebrush	Artemisia tridentata tridentata
Basin wildrye	Elymus cinereus
Bearded wheatgrass	Agropyron subsecundum
Big blueberry	Vaccinium membranaceum
Biscuitroot	Lomatium
Black sagebrush	Artemisia arbuscula nova
Blue elderberry	Sambucus caerulea
Blue wildrye	Elymus glaucus
Bluebunch wheatgrass	Agropyron spicatum
Bluegrass	Poa
Bottlebrush squirreltail	Sitanion hystrix
Cheatgrass	Bromus tectorum
Cinquefoil	Potentilla
Cluster tarweed	Madia glomerata
Columbia brome	Bromus vulgaris
Common beargrass	Xerophyllum tenax
Common camas	Camassia quamash
Common chokecherry	Prunus virginiana
Common cowparsnip	Heracleum lanatum
Creambush oceanspray	Holodiscus discolor
Cutleaf balsamroot	Balsamorhiza macrophylla
Douglas-fir	Pseudotsuga menziesii
Engelmann spruce	Picea engelmannii
Eriogonum	Eriogonum
Geranium	Geranium
Grand fir	Abies grandis
Hairgrass	Deschampsia
Heartleaf arnica	Arnica cordifolia
Hood phlox	Phlox hoodii
Idaho fescue	Festuca idahoensis
Indian ricegrass	Oryzopsis hymenoides
Lodgepole pine	Pinus contorta
Longtube twinflower	Linnaea borealis longiflora
Low Oregon-grape	Berberis repens
Low sagebrush	Artemisia arbuscula
Lupine	Lupinus
Mallow ninebark	Physocarpus malvaceus
Medusahead wildrye	Taeniatherum asperum
Mock azalea	Menziesia ferruginea
Mountain big sagebrush	Artemisia tridentata vaseyana
Mountain brome	Bromus marginatus
Myrtle pachystima	Pachystima myrsinites
Narrowleaf pussytoes	Antennaria stenophylla
Needleandthread	Stipa comata
Nevada bluegrass	Poa nevadensis
Penstemon	Penstemon
Pepperweed	Lepidium
Pine reedgrass	Calamagrostis rubescens
Plains pricklypear	Opuntia polyacantha
Ponderosa pine	Pinus ponderosa
Prairie junegrass	Koeleria cristata
Pyramid spirea	Spiraea pyramidata
Quaking aspen	Populus tremuloides

Queencup beadlily	Clintonia uniflora
Rabbitbrush	Chrysothamnus
Redosier dogwood	Cornus stolonifera
Rose	Rosa
Rough fescue	Festuca scrabrella
Rush	Juncus
Sandberg bluegrass	Poa secunda
Saskatoon serviceberry	Amelanchier alnifolia
Sedge	Carex
Shadscale	Atriplex confertifolia
Silver sagebrush	Artemisia cana
Slender wheatgrass	Agropyron caninum majus latiglume
Snowberry	Symphoricarpos
Snowbrush ceanothus	Ceanothus velutinus
Starry false-Solomons-seal	Smilacina stellata
Sticky geranium	Geranium viscosissimum
Subalpine big sagebrush	Artemisia tridentata vaseyana spiciformis
Subalpine fir	Abies lasiocarpa
Tapertip hawksbeard	Crepis acuminata
Thickspike wheatgrass	Agropyron dasystachyum
Thinleaf alder	Alnus tenuifolia
Threetip sagebrush	Artemisia tripartita
Thurber needlegrass	Stipa thurberiana
Tufted hairgrass	Deschampsia caespitosa
Valeria	Valeriana
Western larch	Larix occidentalis
Western meadowrue	Thalictrum occidentale palousense
Western needlegrass	Stipa occidentalis
Western redcedar	Thuja plicata
Western wheatgrass	Agropyron smithii
Western white pine	Pinus monticola
Western yarrow	Achillea millefolium lanulosa
White spirea	Spiraea betulifolia
Whiteflower rhododendron	Rhododendron albiflorum
Willow	Salix
Winterfat	Eurotia lanata
Wyoming big sagebrush	Artemisia tridentata wyomingensis
Xeric big sagebrush	Artemisia tridentata xericensis

LAND USE

MAJOR TYPE	SOIL
Nonirrigated cropland	Soil 18 (Pywell Series)
	Soil 30 (Oxford Series)
	Soil 35 (Nez Perce Series)
	Soil 50 (Palouse Series)
Irrigated cropland	Soil 10 (Owyhee Series)
Barren land and irrigated cropland	Soil 6 (Sebree Series)
Hay and pasture	Soil 23 (Roseberry Series)
	Soil 47 (Westlake Series)
Rangeland	Soil 12
	Soil 14 (Flybow Series)
	Soil 20 (Bluehill Series)
	Soil 22 (Moonville Series)
	Soil 28 (Moonville Variant)
	Soil 37
	Soil 39
	Soil 41 (Tannahill Series)
	Soil 43 (Gwin Series)
	Soil 48 (Hymas Series)
	Soil 49 (Ola Series)
	Soil 53 (Ager Series)
	Soil 54 (Boulder Lake Series)
Rangeland and nonirrigated and irrigated cropland	Soil 34 (Houk Series)
	Soil 36 (Driggs Variant)
	Soil 42 (Gem Series)
	Soil 45 (Little Wood Series)
	Soil 46 (Rexburg Series)
	Soil 52 (Magic Series)
Rangeland and irrigated cropland	Soil 4 (Chilcott Series)
	Soil 5 (Colthorp Series)
	Soil 7 (Gooding Series)
	Soil 8 (Portneuf Series)
	Soil 9 (Trevino Series)
	Soil 11 (Minidoka Series)
	Soil 13 (Garbutt Series)
	Soil 16 (Quincy Series)
Woodland, hay, and pasture	Soil 31 (McCall Series)
	Soil 44 (Klickson Series)
Woodland and nonirrigated cropland	Soil 2 (Santa Series)
	Soil 3 (Porthill Series)
	Soil 33 (Southwick Series)
	Soil 38 (Greys Series)
Woodland and nonirrigated and irrigated cropla	Soil 29 (Bonner Series)
Woodland and watershed	Soil 1
	Soil 15 (Pyle Series)
	Soil 17 (Shellrock Series)
	Soil 19 (Vay Series)
	Soil 21
	Soil 24
	Soil 25
	Soil 26
	Soil 27
	Soil 40 (Pavohroo Series)
	Soil 51
Watershed	Soil 32